MENSA®

LATERAL THINKING
& LOGICAL DEDUCTION

TEST YOUR POWERS OF THINKING

Dave Chatten
Carolyn Skitt

CARLTON
BOOKS

What is Mensa?

Mensa is the international society for people with a high IQ.
We have more than 100,000 members in over 40 countries worldwide.

The society's aims are:
> to identify and foster human intelligence for the benefit of humanity
> to encourage research in the nature, characteristics, and uses of intelligence
> to provide a stimulating intellectual and social environment for its members

Anyone with an IQ score in the top two per cent of the population is eligible to become a member of Mensa – are you the 'one in 50' we've been looking for?

Mensa membership offers an excellent range of benefits:
> Networking and social activities nationally and around the world
> Special Interest Groups – hundreds of chances to pursue your hobbies
>   and interests – from art to zoology!
> Monthly members' magazine and regional newsletters
> Local meetings – from games challenges to food and drink
> National and international weekend gatherings and conferences
> Intellectually stimulating lectures and seminars
> Access to the worldwide SIGHT network for travellers and hosts

For more information about Mensa: www.mensa.org, or

British Mensa Ltd.,
St John's House,
St John's Square,
Wolverhampton
WV2 4AH
Telephone: +44 (0) 1902 772771
E-mail: enquiries@mensa.org.uk
www.mensa.org.uk

CONTENTS

American Mensa Ltd... 4

Lateral Thinking Puzzles 7

Answers 123

Logical Deduction Puzzles 129

Answers 246

Racetrack Confusion

On the second row of a racetrack starting grid, the driver of car number 7 was the son of the driver in car 3. They had both clocked the same third-fastest qualifying time. The driver in car 3 was not the father of the driver in car 7. How was this possible if he was also not the father-in-law or natural father?

See answer 1

Arise

A man was exploring a mountain when he slipped and fell. He was 150ft from the summit when he slipped, but he was at the top after the slip. He did not climb the rest of the way and he was not lifted to the top by colleagues. How did he slip to the top?

Clues

1. He was on the same mountain and the top was above him.
2. He was not supported by a balloon filled with hydrogen or helium.
3. No ropes or pulleys were involved.
4. No thermals were involved.

See answer 68

High Days and Holidays

 King Henry wanted to change all of the high days and holidays and called his ministers together. He decreed that holidays would occur on the highest day and on the lowest day of each week, and these turned out to be Saturday and Friday respectively. If his week was in the order of high days descending to the lowest day, what would the sequence be?

See answer 38

The Share-Out

Three children were counting the money that they had when they found that they each had only one value of coin. Each child had a different value of coin and each had different numbers of coins. They calculated that if each child gave two of their coins to each of the other two children, they would all have the same amount of money.

If they finished with $1.80 each, how many of each coin did each child have to start with?

See answer 103

My Homework is Right!

At a local infant school a teacher gave the children a few math problems for homework. The next day the teacher pulled Tom out and told him that he had all of his wrong.
His answers to the problems set were:

$$10 + 7 = 5$$
$$9 + 6 = 3$$
$$11 + 5 = 4$$
$$8 + 11 = 7$$

Tom was also right. How was this so?

See answer 90

Bob the Miser's Last Will

Old Bob was a miserly man who never spent his money. His 'Last Will and Testament' stated that he wished to be cremated together with the proceeds of his estate. He did not wish to give his money to his relatives.

When the will was read, the relatives stated that Bob was not sane when he made the will. The judge ruled that he was and Bob's wishes should be followed.

The Judge did, however, find a way to comply with Bob's wishes and at the same time please the relatives. How was this done?

See answer 109

A-Haunting We Will Go!

An ancient castle had been converted into a hotel. After a few months, many ghostly sightings had been reported. The manager was under pressure as many bookings were being lost, but he was getting some business from ghost hunters. The problem was that he could not guarantee to match the appearances with the right guests, until one day he noticed a pattern in the sightings and their timings. If he could predict where and when the ghost would appear, he would keep all of his guests happy.

He found that January through March, room number 3 was haunted every other night. April through June, room number 4 was haunted every third night. July through September, room number 9 was visited by a ghost every fourth night. He then needed to plan which room would be visited in the last quarter of the year and the frequency. How did he work this out and what was his answer?

See answer 40

Uneasy Peace

The warring clans of the Campbells and the McPhersons were brought together by a marriage between the son and daughter of the opposing leading factions. The clan members, however, were still very patriotic to their own clan and were very suspicious of the opposing clan. For the first few years all activities between the clans had an equal number from each clan in the teams of workers. This covered building homes, hunting, fishing, cooking etc.

On one fateful day the fishing boat, which had a crew of 30 (15 from each clan, and headed by the Campbell leader), ran into a very bad storm and the boat began to sink. The head of the expedition agreed with the crew that half of them would have to take a risk and swim for shore in order to save the boat and the remaining crew. The head man said that he would be fair in the selection of those to leave, and that he would line everyone up in a single line formed in a circle and every ninth person would have to go. The crew agreed, and each was allotted a position numbered 1 through 30.

How did he line them up so that only the McPhersons were left?

See answer 2

The Strong Swimmer

A good swimmer jumped from his boat in the middle of the Mediterranean Sea. He swam only 100 feet from his boat and then he sank and drowned. What caused this?

Clues

1. He did not have cramp or any physical or mental health problems.
2. The waves were very light and had no bearing on the tragedy.
3. No third party was involved and his death was not because of an attack by sharks, pirates, etc.
4. He did not get tangled in any nets or weeds.
5. No other swimmer would have survived in his place.
6. The water may have been a few degrees warmer where he sank.

See answer 45

Brother Simon

Brother Simon was a monk of an order that no longer exists. He does, however, have a new job, which ensures that the old monastery collects thousands of dollars each year from tourists. After tourists are shown into his old cell the doors are locked with all of the tourists still in the cell. There is a small window, which is too small to get through, but Brother Simon manages to get out every time. How does he do it?

Clues

1. He does not have a key and the lock is not picked. The door is not opened.
2. The walls are solid with no loose stones.
3. He does not go up or down to escape the room.
4. The room warms up when he leaves the room.
5. I would not go into the room with Brother Simon.

See answer 19

The King is in his All-Together!

We have all heard the song about the king and the magic clothes that only the most intelligent people could see, but did you know that this has since been tested in the opposite direction? A crowd of people who watched a parade saw all of the people in the parade without clothes on. They were all wearing clothes at the time. How was this accomplished?

Clues

1. No hypnosis.
2. No tricks of light or use of special glasses.
3. No use of x-rays.
4. The crowd were not related to Superman.
5. They did not undress or pass by twice.

See answer 82

Lottery Winners

This week's lottery was won by a syndicate of 10 people. Between them they won $2,775,000. They all contributed different amounts into the syndicate and their winnings were calculated against their contributions. If the amounts were all different but the cash differences between each step remained uniform, what amount did the second-highest winner get given that the sum of the lowest three amounts was equal to the sum of the top two amounts?

See answer 67

Antony & Cleopatra

One of the guards at a Roman estate found Antony and Cleopatra dead a few feet away from each other at the end of the day. He immediately called Caesar who confirmed that they were both dead. Caesar ruled out poison and there was no sign of foul play. He did see a small crack in the floor, which ran between the two bodies, and concluded that the crack caused their death. He was right, but how did they die?

Clues

1. They had not been strangled or suffocated.
2. They were both naked.
3. They were both good swimmers.
4. They did not injure themselves by diving into an empty bath or swimming pool.

See answer 36

Nylon Ball-Bearings

A factory produced millions of tiny ball-bearings made from a nylon polymer. These were incredibly tough and light in weight. This made them very cheap. When a few dozen were put on a concrete floor they could support a truck without failing to work as they should. They were stored in very large wood compartments, which were 15 feet deep. Under normal circumstances the material used is quite safe and not poisonous. The death of a worker was viewed rather differently by the coroner. Why?

Clues

1. He was not killed in the manufacturing process.
2. He did not fall as a result of ball-bearings being used.
3. Nothing hit him or crushed him.
4. His death was not caused by toxic fumes from the material or as a result of fire.

See answer 6

Levitating Balloons?

A family had inflated several different-sized balloons with air and tied the ends so that they would not deflate. These were left all over their front-room floor before they went out shopping. When they returned and looked into their front room from outside the house, they were surprised to see that all of the balloons were two inches above the floor. Why was this?

Clues

1. The room and the balloons were at the same temperature and all of the doors were firmly closed. The doors had draft excluders fitted.
2. The balloons did not contain any gases that were lighter than air.
3. The balloons were not held up by strings and no shock wave was involved.
4. Static electricity or electric charges were not the cause.
5. The cause was not air circulation.

See answer 28

Trackside Jo

Trackside Jo had been taken into hospital for a serious heart condition. The nurse who looked after him noticed that he had several betting slips in his pocket when he was admitted but she thought that these should be kept from him until he was well. The extra stress, she thought, might upset his recovery. After two weeks of total rest following his operation, the nurse gave him the daily newspaper and gave him his betting slips and wallet. Looking at his first betting slip and newspaper, he noted that his first horse had won at 50-to-1 and he had $50 to win on it. When he left hospital his first call was to collect his winnings of $2500. They refused to pay him, but do you know why?

Clues

1. There was no time restriction on the betting slip.
2. The bet was valid and he had paid $50.
3. The bookmaker had not disappeared or gone bust.
4. He did not owe $2500, or more, to the bookmaker.
5. He had not made a mistake when filling out his betting slip.
6. The horse had won and was not subject to disqualification.

See answer 55

No Fire for Explorers

Neil and Dave were exploring a new territory. They felt a little cold and decided it was time to build a fire from some newspapers and dry bits of wood, which they had brought with them. The matches were all unused and dry, but would not light the wooden part of the match; their lighters would not work even though they seemed as though they should, and they even resorted to flintsparks and using the sun's powerful rays through a strong magnifying glass.
Nothing worked. Why?

Clues

1. They were above ground and in the open.
2. There were no winds or draughts.
3. It was not wet or humid.
4. The newspaper was not wet or damp.
5. All of the equipment used to light the fire was in perfect condition.

See answer 98

King-Elect

The king had died some time ago and the queen replaced him on the throne as Head of State. They had two children who were twins. Both were delivered at birth by caesarean section, and both were born at exactly the same time.

A king had to be chosen. One of the two was very intelligent and loved by everyone, but the other was not so bright. He was not liked at all. and was not favored by the queen or people in their parliament. It was the latter who was chosen. Can you work out why?

Clues

1. There were no corrupt motives involved.
2. The constitution was used to elect the king.
3. The intelligent child did not die and was not harmed or locked away.
4. The queen agreed with the decision.
5. Foreign powers were not involved.
6. Marriage did not form part of the decision.

See answer 76

Problems with Air Pollution

A chemical plant had a major fire, which was so ferocious that it lasted 12 hours before the fire department got it under control. The police had to evacuate all of the houses within a one-mile radius because the fumes were so toxic that they would kill anyone who inhaled them within minutes. The wind initially blew the toxic gases from the west toward the east, and the wind blew constantly for 3 hours and 20 minutes. The police, however, started to clear the houses on the west of the plant because this seemed to make a great deal of sense. This evacuation procedure saved thousands of lives, but then the wind changed to blow from the east toward the west. Those who had not been evacuated either died or had serious medical problems. The wind continued to blow in this direction until the fire was completely extinguished. Only those people living to the west of the plant died. Why?

Clues

1. It did not rain.
2. Deadly toxic fumes were released all the time.
3. The toxic fumes were heavier than air and did not go over and beyond the one-mile danger circle in the east.
4. Nobody in the east had breathing apparatus and none was evacuated.
5. Closing doors and windows did not give total protection.

See answer 101

Happy New Year and Again and Again etc.

It is August and a 26-year-old woman said that she had never missed a New Year celebration in her life. She also claimed to have seen "The New Year" in 51 times. How could she be telling the truth if she was born in June?

Clues

1. She only counted January the First as a New Year and other religious or cultural New Years were not counted.
2. She did not cheat by winding her clock back.
3. Her 26 years were using a modern calendar and she lived in modern times on the planet Earth.

See answer 44

Head-On Ant Crash?

A rod of steel has a line painted on it from one end to the other. It is then twisted in the middle so that the line is half on one side and half on the opposite side. The line is painted only as wide as a quarter of an ant's width. These ants are intelligent and they are told that they must remain on the line or perish. The ants are placed at either end of the rod and told to go to the other end of the rod where they will be fed in safety. If they meet each other, they will both be killed. How do the ants resolve this problem and achieve their goals?

Clues

1. The rod was solid and could not be made hollow.
2. The ants could not avoid each other if they were both on the line. They could not jump over each other.
3. Both ants were fed and neither died.
4 The rod was not suspended or spun round.

See answer 8

Sinking Robots

Mission control had calculated everything down to the last detail. Experiments conducted on the planet ZOD on a previous visit showed that the mobile robots would be able to walk on ZOD's surface. The spacecraft was, however, blown off course and was forced to land on a planet similar in size to ZOD. The two robots were ejected before the crash landing and were not damaged in the descent to the surface. They then sank below the level of the surface and could not move. What caused this?

Clues

1. The soil make-up was the same on both planets and the soil density on the crust of both planets was the same.
2. They did not land on wet ground or water.
3. They did not sink because of impact speed on landing.
4 If they had ejected over ZOD they would not have sunk.

See answer 16

St. Joseph's Church

Daniel's family were very religious and always went to church on Sundays. Daniel's father had been asked to relocate because of a job promotion and they moved the family to a new city on the Saturday. The move was difficult and took all day and all of the family, except Daniel, slept in on the Sunday morning. Daniel felt tired but decided he would find his local church and thank God for the safe move straightaway, and he would lead the others to church a little later in the day. The sign outside the church said St Joseph's Catholic Church. He entered to find a service was being conducted but he did not understand a word that was being said. Why?

Clues

1. It was nothing to do with accents.
2. They had not moved to a country outside the USA.
3. Daniel was only 10 years old.
4. The language used in the church was English and not Latin.
5. He did not have a problem with his ears. He could hear everything that was said.
6. The city they moved to was Washington.

See answer 88

The Rejected Recruit

Bright Sam was desperate to work in electronics for the Army. He was one of the brightest people in his class and excelled in electronics theory. When he failed to get into his chosen trade in the Army he was devastated. He knew that he was best qualified yet the Army did not want him. He later received a letter from the Army offering him a job that could save his colleagues' lives, a job that would involve him using his special gift. Can you work out what this was based on the clues given below?

Clues

1. He could not do electronics or signals because he was color-blind.
2. He was physically fit and intelligent.
3. His vision was very good, other than his problem with colors.
4. He was young and ideal for combat.

See answer 5

The Great Soccer Player Retires

A great player who had given his club and country years of good service was honored by being given a testimonial soccer match between his club and his country. It was to be his last game before retirement. The match score was 3-2 and he had scored four goals but finished on the losing team. Can you work out what happened if:

Clues

1. He scored all of his goals at one end of the stadium.
2. The winning goal was an own goal and it had not been scored by him.
3. He turned around at half-time to play in the opposite direction from the way he played in the first half.

See answer 70

The Deadwood Stagecoach

In the days of the Wild West a prospector was planning to go on the morning stagecoach to take him back east. He had struck gold and decided to celebrate in style. That evening he had drunk enough whiskey to make him drunk twice over. In the morning he found himself on the stagecoach, but it would not take him back east. Why was this?

Clues

1. He had a valid ticket and money for the trip.
2. He was on the stagecoach well before anyone else.
3. Other people wanted to go back east but nobody stopped them.
4. There was plenty of room for him on the stagecoach.
5. It was not a dangerous trip.
6. The driver for the stagecoach and the horses left on time.

See answer 34

The Mountaineers

A family of 4 were going on a mountaineering holiday. The second morning they were all found dead in their cabin. The coroner declared that they had all died from drowning. The faucets in the cabin had not been left on and the boiler and water storage units were undamaged. There was no sign of any foul play. What caused them to drown?

Clues

1. They were at least a mile from the nearest lake.
2. It had not rained for 5 days. Not a flash flood.
3. It was not caused by problems with a dam.

See answer 116

Little Breeders

A man went to a pet shop and asked for a pair of budgerigars for breeding purposes. The shop-keeper sold him a pair of birds who seemed inseparable in the shopkeeper's cage. Six months later the man revisited the shop to complain that no eggs had been produced. The shopkeeper wished to keep the customer happy and gave him another budgerigar that had just laid eggs and reared the young. Six months later, the man returned again with a story of disappointing failure. Why did the hen birds fail to lay a fertile egg?

Clues

1. There was nothing wrong with any of the birds.
2. They had the right diet.
3. They were all at an age that was right for breeding.
4 It was a quiet and peaceful house.

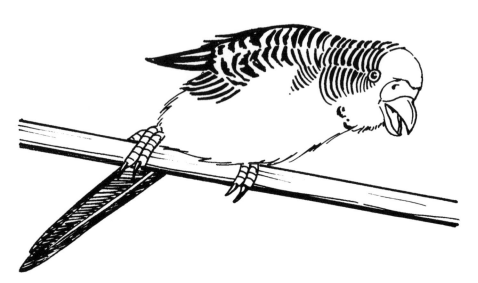

See answer 26

Who is the Bigger Liar?

In a prison a bag of sugar went missing from the kitchen. You are told in their statements that four of the five main suspects have told one lie and the other suspect has told two lies . The culprit is the person who has told two lies. Can you work this out given the following statements:

Prisoner A : I was in my cell with prisoner D and I could see prisoner B in the workshop, prisoner C in the showers, and I could also see prisoner E in the gym from either the cell door or cell window.

Prisoner B : I was in the workshop and I saw prisoner D in the gym. I also saw prisoner E in the gym and prisoner C was having a shower. I did not see prisoner A look at me through his window but he may have been in his cell.

Prisoner C : I was having a shower but I did see prisoners A and D in their cell, and also saw prisoner E in the gym. I could not see the people in the workshop.

Prisoner D : I was not with prisoner A in the cell, but I saw prisoner B in the yard watching prisoner C in the shower, and he was also watching prisoner E working out in the gym.

Prisoner E : I looked in prisoner A's cell and he was not there, but prisoner D was. Prisoner B was in the workshop and prisoner C was in the shower following a workout in the gym with me. He left me to complete my exercises.

See answer 54

Leaky Pipe

A pipe sprung a leak on its underside so that it leaked 5 gallons of water per hour until the pipe was empty 4 hours later. The leak was not detected and the pipe was refilled but a second leak, of exactly the same size, occurred immediately. The pipe was now leaking at a rate of 10 gallons of water per hour but this time it took 3 hours to empty . Can you understand why?

See answer 92

The Bus Drivers

Two bus drivers sit chatting in the staff canteen. One of the drivers leaves the canteen to meet a young boy waiting outside. A third bus driver entering the canteen asks the driver with the young boy who the boy is. "He's my son," replies the bus driver. The third bus driver sits down in the canteen and hears the other driver in there claiming that the boy is his son too. How can this be? The boy does not have any step-parents and both bus drivers are telling the truth.

No Clues : This one is quite easy.

See answer 75

The Bouquet of Flowers

A florist is making up bouquets using roses, carnations, and chrysanthemums. Twice as many of the bouquets contain carnations only as chrysanthemums only. There is one more bouquet containing roses only than carnations only. There is one more bouquet containing all three types of flowers than a mixture of roses and carnations only. There are exactly the same number of bouquets containing roses only as a mixture of carnations and chrysanthemums only. There is one more bouquet containing both roses and chrysanthemums only than containing chrysanthemums only. Two bouquets contain chrysanthemums only and 18 do not contain any chrysanthemums.

Clues

1. How many bouquets contain roses only?
2. How many bouquets contain only two of the three types of flowers?
3. How many bouquets contain carnations only?
4. How many bouquets contain all three types of flowers?
5. How many bouquets does the florist make in total?

See answer 108

Charged by a Bull

Four ramblers walked down the lane, past the stream, over the hills to the edge of a field. The field was full of cattle. Before the ramblers managed to reach the other side of the field they were charged by a bull. Why did the ramblers make a formal complaint when none of them suffered an injury?

Clues

1. They did not run to safety.
2. They were not scared.
3. The cattle took no notice of the ramblers.
4. Bulls had charged others in the past but not for a period of time.
5. The charging bull was fully fit and fully grown.

See answer 48

The Fan

A young boy going to an important soccer match decided to paint his face green, the color of his favorite team. His team won the match and he celebrated for hours with his friends after the game. When he got home he was dismayed to discover that his face was blue and not green. Why?

Clues

1. He was not painted a second time.
2. The paint was not affected by ultraviolet light.
3. The paint did not dry to blue when it was applied.
4. The paintbrush was clean and had no chemicals on it.
5. The change was not a result of temperature-sensitive or light-sensitive additives.

See answer 106

Big Bill

Big Bill was extremely tired one evening so he turned the light off and got into bed. The next morning he awoke to hear on the radio of a terrible tragedy that happened in the early hours of that morning, killing over 100 people, and it was all his fault. Why? He did not wake up and he did not sleepwalk.

Clues

1. The weather outside was bad, with poor visibility.
2. Bill was tired as he had not slept the previous night or through the day.
3. His alarm bell had been broken and it no longer rang.
4. If he had done the same things the night before, more people would probably have died.

See answer 17

Bush Fire

There was a forest fire in Australia. After the firefighters had managed to extinguish the fire, the search for bodies began. After two days of searching they found a man in complete scuba diving gear. Although he was dead, he had not been burned at all. The forest is 20 miles from any water. How did he get there?

Clues

1. The man had not walked to where he was found.
2. The man had not been murdered. It was an accidental death.
3. His wet suit was not burned or melted.
4. The man had several broken bones.

See answer 83

The Arabian Prince's Car

The Arabian prince bought a top-of-the-range car with white leather seats, state-of-the-art hi-fi, television, and every extra imaginable. It was his pride and joy to own such a car. When he got it he found that it had that `new' smell so he stuck an air freshener on to the top of the front windscreen, and it dangled from the sucker by means of a string. After only one hour the perfume from the air freshener gave the car a beautiful smell, and the prince was very happy. He decided to drive to his father's palace in the desert to show him his delight. He left a newspaper on the dashboard and a present on the back seat for his father. A guard was posted to look after the car. His father was out but returned 2 hours later to find his son in the palace waiting for him. He rushed his father into the courtyard to find the car on fire and the guard throwing water over it. What caused the fire?

Clues

1. The fire started inside the passenger section of the car.
2. No electrical or fuel problems existed.
3. The present did not contain any flammable materials.
4. Firearms and matches were not involved.
5. Spontaneous combustion was not the cause.
6. It had nothing to do with chemicals in the air freshener.
7. The guard had nothing to do with the cause of the fire.

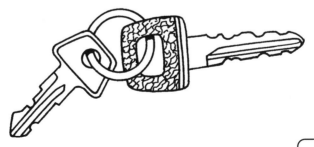

See answer 64

Moving Suitcases

A family on vacation in Florida returned to their rented apartment one day, only to see their empty suitcases placed at the side of the road almost 2 miles from the apartment. They stopped their car and inspected the cases, which had their names on the name labels. Why had their suitcases been placed at the side of the road?

Clues

1. They had paid their rent and still had one week's rent paid in advance.
2. They had been away for the day.
3. They had not been burgled.
4. The landlords did not have the suitcases removed.
5. They had left the suitcases in the apartment before leaving for the day.

See answer 37

Amateur Safe-Crackers

Two cowboys, Lightfingers Harry and Desperate Dave, decided that they would blow open the safe in the town bank, which contained many thousands of dollars. They had never blown a safe before but they knew where they could get as much gunpowder as they might need. Over a drink in the saloon they asked a drunk gold prospector how much gunpowder they would need. He told them about 2 pounds, but Desperate Dave insisted that they used twice the amount to make sure. They entered the bank, poured the gunpowder, and lit the fuse. The safe did not open and not a sound was heard. Why?

Clues

1. All of the gunpowder was used and it ignited. The powder was dry and they used all 4 pounds.
2. They did not try to soundproof the room and it was not soundproofed already. People were nearby.
3. The powder was placed on and around the safe, close enough to get the job done.

See answer 41

The Immovable Screw

A man decided to repair his wife's vacuum cleaner (much to her despair, since he had shown no aptitude with electrical or mechanical problems in the past). The first job was to remove the screws using his screwdriver. He ensured that the right size and type of point on the screwdriver matched the screw head perfectly. He then engaged the screwdriver to the screw head, applied the necessary force, and turned the handle counter-clockwise. The screw would not come out and it would not loosen. Why?

Clues

1. The pressure applied by the husband was adequate to remove the screw.
2. Turning the screw counter-clockwise was the correct way to loosen and remove the screw.
3. Good contact between the screw and screwdriver was maintained. The screwdriver did not slip off the screw head.
4. The thread in the hole did not get stripped and the screw was not damaged or deformed.
5. His wife was able to undo the screws without any problems using the same screwdriver and without lubricants.

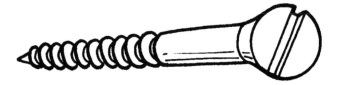

See answer 27

The Tea Party

A mother calls her daughter to come and play in the house. The little girl comes running through the front door and decides to have a tea party with her dolls and teddy bears. After half an hour she is bored with this game, and decides to go back outside to play with her ball in the front garden. To get to the front garden she has to go through two front doors. Why?

Clues

1. The house does not have a porch door.
2. One front door is facing the back wall of the house.

See answer 56

Two Brothers

In 1914 there were two brothers of an aristocratic family in England. When war broke out the first brother volunteered for the army without delay. After basic training he was sent to the front line. He was an officer and led his unit with complete distinction for over 12 months. Upon his return for a rest he went to his family home to find his brother just having a good time. For generations his family had served their country with honor, but his brother was bringing shame on the family name. The officer returned to the front and suffered an injury; while in the hospital he sent a letter to his brother, which caused his brother to enlist as a foot-soldier and win medals of distinction and bravery. The letter sent did not contain a letter or any words from his brother. Yet he knew by what was in there what it meant. What was in the letter?

Clues

1. The handwriting on the envelope was not his brother's.
2. It had a postmark that could not be read.
3. His brother did not speak to him to cause him to change his mind.
4. The envelope contained something that weighed no more than the envelope itself.
5. There was no message on the envelope.

See answer 95

The Bath of Liquid

A man fell into a full bath of liquid at work. When he got out he was dry, but he was taken straight to a hospital. Can you explain why he was dry and why he was taken to hospital?

Clues

1. The liquid in the bath was at room temperature.
2. There were warning signs to keep clear.
3. It was an accident that caused him to fall into the bath.
4. He had fallen gently and had not suffered a concussion or any severe blows.
5. The liquid in the bath was 4 feet deep, and little was lost when he fell in.
6. He was not wearing any protective clothing.
7. He had not ingested any of the liquid.
8. He was required to burn his clothes.

See answer 77

Car Park Overcrowding

A company had a car park where all of the 10 spaces were allocated to its managers. They expanded the business and a new manager joined them. Part of his contract was to have a car-park space, just like the other managers. How was this achieved if nobody was asked to double-park?

Clues

1. The cars could not obstruct either of the access roads.
2. All of the spaces between the cars had to remain the same.
3. The extra car could not be parked in a location away from the front office wall, and all of the other managers kept their slot.
4. All of the cars needed to be parked at the same time.

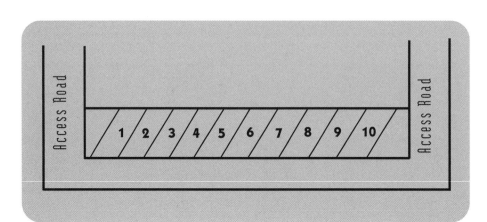

See answer 109

The Courier's Wait

The courier phoned his customer to say that the crate that he had brought with him weighed one ton, and that they would need lifting equipment to unload it. He was less than a mile from the delivery point but it would be 6 hours before he could get there. He had covered the 20 miles from the collection point in just over one hour. Why would it take so much longer to reach the delivery point given the following clues?

Clues

1. He was not taking a detour, and there was no traffic between his current position and the delivery point.
2. He was not being held up because of other meetings or people.
3. The delay was not caused by unloading or loading any other products.
4. If it was $5^3/_4$ hours later, he could make the same journey in 15 minutes.
5. The roads in the area were free from traffic congestion and road works.
6. The reason was not due to anything anyone did.

See answer 23

Leap to Safety

A man sleeping on the top floor of a three-storey house awakes to find smoke coming under his bedroom door. He gathers as many of his treasured possessions as he can possibly hold and leaps out of the bedroom window. Even though his arms are full, he doesn't drop or break anything and he does not injure himself. Why?

Clues

1. Some of the items were fragile and would have broken if they had hit the ground.
2. He did not jump on to a ledge on the house.
3. No ladders, ropes, or safety nets were employed.
4. He did not jump into water or soft snow.

See answer 50

The Class

James trudges off to school each morning with his books but he rarely does homework, and he doesn't achieve high marks in tests either. There are 36 children in his class and 35 of them are good students. Why does James never get into trouble?

Clues

1. James is always polite.
2. James has been sent to the head's office on a number of occasions.
3. James is not related to anyone at the school and he is not a special student.

See answer 20

A Fruity Problem

A woman has a small collection of artificial fruits on the windowsill. The apple is rosy red on one side and bright green on the other side, and there is a little white stalk sticking out from the top. The peach is a lovely soft warm shade of pinky-orange, with a larger white stalk. There is also a pear and a deep burgundy-colored plum. The woman leaves them on the windowsill and goes out of the room. When she returns half an hour later she cannot see the fruits at all. Why? Nobody else has been in the room. Nobody has moved them. There is nothing blocking her view; the room is clear, and there is no mist.

Clues

1. They had not been stolen.
2. They had not been eaten.
3. A telltale clue had been left.
4. The room had an unusual smell about it when she returned.
5. Animals and insects had nothing to do with the disappearance.

See answer 85

Disappearing Treat

At a candy store a young boy was allowed to choose what he wanted. He came out of the shop happily clutching a full bag. He made a hole in the top of the bag and began eating. He only ate a small amount of the contents but within half an hour his bag was virtually empty. He did not drop the bag or its contents. He did not give any away, throw any away, or transfer the contents into anything else. Where did the contents of the bag go?

Clues
1. Only about 5% of the content of the bag had been consumed.
2. The hole in the bag did not let any of the contents out.
3. The contents were not eaten by insects or anything else.

See answer 39

The Messy Eater

Much to his colleagues' annoyance, Arthur brought fruit to the office each day for his lunch. He would peel his banana and leave the skin lying around, drop apple cores all over the floor, spit the pits from his grapes over other peoples' desks, and he was forever squirting people in the eye with his orange. Arthur still brings fruit to work but no longer gets complaints from his colleagues. He has not changed his habits, he has not done or said anything to his colleagues, and his colleagues have not changed. Why does he no longer get complaints?

Clues

1. He did not work with animals and the office was normally a clean environment.
2. He no longer used his fingers to hold the fruit.

See answer 63

The Fabric Shop

In a curtain shop there are flowered fabrics hanging up in the section marked `Floral Designs`. All the curtains in various colors but with no pattern on them are in the section marked `Plain Fabrics`. Why are a pair of curtains with continuous vertical lines down them not in the section marked `Striped Fabrics`?

Clues

1. There was a section marked `Striped Fabrics`.
2. They were vertical stripes.
3. They had not been misplaced in another section.
4. The customers knew where to find the curtains they needed.

See answer 35

The Disappearing Man

One cold winter morning Jayne was walking down a narrow country lane. On either side of the lane there were four houses. Jayne noticed that each house had a different-colored front door and different-colored cars parked in the driveways. Outside one of the houses she noticed a man standing in the garden. He was very well dressed with a hat and scarf on to keep him warm. She waved at the gentleman and shouted, "Hello!" and he smiled at her. Later that day when she came back along the lane she noticed the man again. She waved to him and said, "It certainly is getting warmer, it doesn't feel as cold as it was this morning." The gentleman smiled at her and she went on her way, counting the cars that passed her as she went. The next day when Jayne went down the lane she noticed the gentleman had gone. Where?

Clues

1. He had not gone inside the house or any other house.
2. He had not walked down the lane in any direction.
3. He had not driven anywhere by car.

See answer 7

Washing Dishes

A married couple in New York had six children and each night of the week one child would wash the dishes. This task was performed by a different child every night. On Sundays all of the children would draw lots to see who would have the sad privilege. One of the children figured that it was best to be left the last lot and not pick at all. She calculated that the first pick would have a 1 : 5 choice, the next a 1 : 4 choice, the next 1 : 3, etc until she was left with the last lot. The child added all of the previous factors together and decided that it would be unlikely that it would be the worst lot left. Was this trick likely to work?

See answer 29

1930s

On one early transatlantic flight in the 1930s a plane carrying 20 passengers had very low fuel reserves when approaching New York from England. It was a very windy day when the plane arrived, but it could not land where it was supposed to because of the wind. It was, however, able to land only a few miles away where the wind had a slightly higher speed. Why was this possible?

Clues

1. The wind direction for landing at the second landing point was less favorable. It was more of a crosswind than the wind at the first landing point.
2. The first arrival area had no other vehicles on it and no other air traffic was involved.
3. Air traffic control did not advise of anything being wrong with the plane, and indeed nothing was wrong with the plane.
4. The plane was not diverted because of the low fuel situation.
5. The pilot could see why he should divert the plane.

See answer 57

Dangerous Neighbors?

The Price family were regarded by their neighbors in Quietsville as complete undesirables. At least one of the family would always be terrorizing some neighbor. The neighbors were too frightened to speak to the police because of their fear of reprisals based on a long history of previous events. One day the situation escalated into a much more serious problem when one of the Price family set fire to a neighbor's home. The police questioned all of the neighbors, but even though some knew who did the deed they would not say. One neighbor handed a note to a policeman, and he went straight to the right member of the family. If the family names were Mr Tom Price (father), Mrs Julie Price (mother), and the children were James, David, Mark, and Chuck, which family member was arrested?

See answer 96

The Last Train

A man went to the railway station to catch the 12:47 train. When he arrived, he realized that he was not wearing his watch. As he walked past the ticket machines, he saw a clock. The man then thought that he was an hour and a half early, so he walked away from the platform. A short while later, he realized his mistake when he missed the train. The clock was correct, so why did the man think he had been early?

Clues

1. His watch was showing the correct time.
2. He did not ask anyone any questions.
3. He had not read anything about delays.
4. His train was on time and had not been re-scheduled.

See answer 74

The Unlucky Locksmith

A locksmith was called to an exclusive bank and asked to change the lock on a room that was used to store valuable documents. The door was to be activated only by the breaking of two low-power laser beams in front of the door. This would release a steel plate that covered the lock, and the owner could then use the new special key to open the lock. The system was to be automatic and re-set itself after use. Just before he had completed his clearing up, the manager of the bank wanted to check it out and after helping the locksmith to clear his tools from the storeroom, he was locked into the storeroom. The locksmith could not get him out. Why?

Clues

1. The door had closed by accident or by design.
2. The police and fire department had to release the manager.
3. The locksmith had to change the lock again.
4. The locksmith still had the key but he could not make it work, even though he had tested it before clearing up.
5. The laser beams were only 3 feet apart.

See answer 105

The Glass Head

In recognition of the President's services to his country and for his contribution to world peace, a huge, two-ton polished glass head of his likeness was commissioned, the base of which was to be flat to ensure that it did not move on its plinth. The top of the plinth matched the neck of the glass head perfectly. An overhead crane with specially padded ropes was used to lift the head on to the neck, but then a problem occurred. The two parts had to be positioned exactly, and the workers could not drag the ropes, since this would chip the head or base. How did they do it?

Clues

1. They could not use wooden wedges or anything that might scratch the glass.
2. They could not use compressed air since the compressor did not have the power.
3. The ropes had to pass under the neck in 4 places.
4. The ropes were made of nylon, which covered a stainless steel core. They were 2 inches in diameter.
5. They could not use suckers or rubber props.

See answer 115

Survival of the Weakest?

Three men were passionately in love with a lady. The woman loved them all equally, but the passion boiled over and the men agreed that they should have a duel using pistols. To the victor, the hand of the lady; to the vanquished and defeated, death, injury, or disappointment.

After agreeing to duel, the odds were stacked against one and in favor of the other two. As Count Nevermiss was an expert and a perfect shot, he had won every duel even against better opposition than he was to face that fateful day. Lord Bullseye was a good shot and a military man. He could be relied on to hit his target two out of every three shots, while Captain Missalot could only be relied upon to hit his target once every three shots. They were, however, men of honor and decided that the rules of the duel gave the poorer shots a chance. They decided that they would stand and face each other from three points of a triangle. There was no limit to ammunition, but they would shoot in turn at either of the opponents with the worst shot going first and the best shot shooting last.

You are put in Captain Missalot's position. How do you maximize your chances of survival with honor? It is you who will shoot first. Who are you going to go for? Survival depends on good lateral and deductive reasoning.

See answer 42

A Problem for the Ferryman

A man leaves his 5 children with the ferryman and is told that they must all be taken to the other side of the river in a minimum number of crossings, such that each of the children has an identical number of one-way trips. The children are all of different ages and the ferryman can only take himself plus a maximum of 2 children at any time. No pair of children of neighboring ages can be left in the absence of the ferryman. Only the ferryman can row the boat. How many trips are needed and what is the sequence?

See answer 18

Household Enquiry

A man enters his house. He asks his daughter, Sally, a question (to which he does not know the answer). The question is such that whatever the answer, right, wrong, true or false, he will know the answer to his question. What was the question?

Clues
1. His daughter did not have prior knowledge of the question.
2. She could have answered with any word or words.
3. She was not in the room when the question was asked.

See answer 84

Long ago in the days of the Pharaohs the Puzzle King was a very favored man – so much so that one of the Pharaohs allowed him to design the entrance to his tomb. The Pharaoh said that his tomb must not be plundered after his death so the design had to deter his people from trying to enter. He would also have 200 of his strongest soldiers entombed with him in case he revived and needed to be released. The design for the entrance is shown below. The magic cube would seal the entrance. How did the cube go together before it was moved into position in the pyramid so that the Pharaoh could get out?

Clues

1. The cube was solid and made from stone. It was made in two halves as shown.

2. Dovetail joints were on the faces that you cannot see and are in central positions. The cube looks the same from each side view. Each side has the same dovetail joint showing.

3. 200 men could move half of the cube but they would not be able to move the entire cube. It took 400 men to move the cube into place.

4. No outside help was needed.

5. No hinges or tricks were employed.

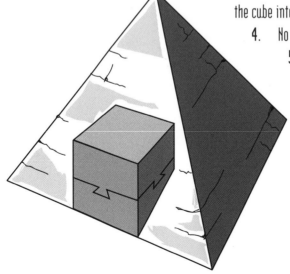

See answer 66

Target Practice

The twins Larry and Pete got up one morning and painted some large targets on the door of the barn. After the paint had dried, they found that their baseball damaged the door if they pitched a fastball. The rubber ball and tennis balls were either lost or of no use because they did not leave a mark to show where they hit the targets. The twins, who were very competitive, did have a solution, one that also pleased their parents. They threw balls at the door for hours and could accurately score every shot without leaving a mess to be cleared away later and without damaging the paintwork. How was this possible?

Clues

1. The balls had no dye and no mud.
2. The balls did not bounce.
3. The twins were told to clear the yard before they could play ball. This instruction was in their best interest.
4. The children kept clean.

See answer 32

Sally's Wash

Sally goes to the bathroom to have a wash. She wants to run a full basin of water so that she can get a nice lather on the soap, but unfortunately the plug for the basin has been lost. She cannot find another plug anywhere and cannot find anything else to fill the plughole. She knows, however, that while the water from one faucet will not stay in the basin, the water from the other faucet will not run away. Why is this?

Clues

1. She does not jam the soap in the plughole.
2. The plughole can let out water more quickly than both faucets on full.
3. A few days before and a few days later she could not have used this idea.
4. She had to run the other faucet to clean the basin.

See answer 3

Hold-Up Clues

A man walks into a bar and asks for a glass of water. The barman goes into the back and then returns to the bar wearing a mask and holding a gun. The customer thanks him and walks out of the bar without ever having any water. Why was he satisfied?

Clues

1. The barman did not know the customer.
2. The customer was not a criminal.
3. The barman did not give or take anything from the customer, although the customer lost something.

See answer 25

The Aircraft

Why did the men fill the transatlantic passenger jet's fusilage with water?

Clues

1. It was safe to do so.
2. The jet was not on fire or a fire risk.
3. Passengers were at risk prior to this being done.
4. It was not an emergency procedure after landing on water.
5. It was not a safety drill.

The Inherited House

Jamie did not know his uncle had left him Sea View House when he passed away. He knew that it was a mansion built about 200 yards back from the cliffs overlooking the sea. He had been there when he was a child, some 30 years ago. Jamie was not close to his uncle, but he was the last surviving relative. It had taken legal investigators some years to find Jamie, as he worked overseas. When he saw the mansion again, he was very disappointed. Why?

Clues

1. It had been well maintained and was in good order.
2. No building had been placed between the house and the sea.
3. The gardens were still in good order.
4. The nearby town had prospered.
5. It was not sentimental disappointment.
6. His uncle had lived there until he died.

See answer 93

A Bargain

Why did the multi-millionaire decide to buy land that was over 200 yards from the seashore?

Clues

1. It was under the sea.
2. It did not contain any mineral rights and it had nothing to do with mining.
3. There was no oil for hundreds of miles.
4. It was not a port or going to be a port or harbor.
5. It had nothing to do with swimming rights.
6. It was a bargain.

See answer 71

The Fire Drill

At a school in Florida the fire bell sounded for a fire drill. The children and teachers were orderly and knew what to do. The children did not know that it was a practice session. The fire department, however, were needed because a major state of panic ensued. What occurred?

Clues

1. The teachers and children could not exit the building.
2. The Fire Department knew of the fire drill but they were not required to be on site for the drill. They were, however, summoned.
3. Many lives would be saved by not leaving the school building.
4. Fire was not involved.
5. Doors and windows were closed throughout the school.

See answer 107

Don't Jump to Conclusions

A man was born before his father and he married his three sisters. He did nothing against the laws of God or man. How was this so?

Clues

1. He remained celibate all his life.
2. He only had one father but worshipped another.
3. He did not belong to a religious order that permitted close family or multiple marriages.
4. His father was 30 years older than he.

See answer 49

The Musical End

The entire family gathered around Grandpa's bed in the hospital at visiting time as usual. He had been in a coma for a few days but he was not expected to die in the near future. The piped music in the hospital stopped suddenly and Grandpa died almost immediately. Why?

Clues

1. He was hooked up to monitors and drip feeds.
2. He was hooked up to life-support equipment.
3. The equipment did not fail.
4. The power supply to the equipment did not fail.
5. His death was preventable.

See answer 13

A Body Bag in the Suitcase

Cheryl had just met a new boyfriend, Floyd. They met in Las Vegas and got married after a whirlwind romance. When they loaded up the car she looked into a suitcase that she had not packed, which had been left in the trunk of the car by her new husband. It contained a body bag with a boy's body in it. The suitcase had holes in it so that air could get into it and the body bag was partially open. She did not leave Floyd or report the incident to the police. Why?

Clues

1. He told her that she had found his best friend.
2. The body bag was used for protection.
3. The boy was 7 years old.
4. It was not his son.
5. The body was fully dressed.
6. Foul play was not suspected even though an arm had been broken.

See answer 89

Cardinal Lock'emup

The Cardinal was given the king's own writing desk for catching and locking up a musketeer who, it was alleged, seduced the queen. The desk was magnificent, with thousands of inlays and studded with jewels. It had 4 crystal inkwells and a drawer for 20 quills. The Cardinal, who was the Minister for Justice, knew that the musketeer was not guilty, but it suited his plans to have him executed. On the day before the scheduled execution, 3 musketeers had an audience to plead with the cardinal for leniency. The Cardinal would not listen so the musketeers made him listen at the end of a sword and made him write out a release paper, which is shown below.

> To The Captain of Guards
>
> I authorize the immediate release
> of Musketeer Antonio.
> He is innocent of the charges made against him
>
> Signed : Cardinal Lockemup

Clues

1. The musketeers saw the letter being written.
2. The format and seals for the letter were in order.
3. The Cardinal had not anticipated such a move and had not given special instructions to the Captain of Guards.
4. The king and queen did not know what was going on.
5. The musketeers did not have an arrest warrant out for them.
6. Antonio was the right musketeer and they went to the correct jail.
7. The Cardinal did not raise the alarm.

The letter was sealed using the Cardinal's seal, rolled up and sealed again. The Cardinal was then asked to ensure that he would not be disturbed for 2 hours. He was then bound, gagged, and locked in his room. The 3 musketeers then went to collect the other musketeer and were all arrested. Why?

See answer 61

The Burglar

A burglar climbed into a house but unusually for him, made a large amount of noise. He ran through the house and identified the owner's most precious treasures and ran out of the house with them. On exiting the house he found that the police were already waiting for him. The homeowners did not press charges and the police took the case no further. The neighbors who had been awakened by the commotion, however, insisted that the man be arrested. What was going on?

Clues

1. The house was alarmed.
2. The noise made by the burglar woke everyone in the house.
3. He had to get out of the house quickly.
4. He was planning to burgle the house.
5. The judge was lenient.

See answer 33

The Policewoman

The policewoman just watched as a man tried to pick a lock to enter a house. He failed to get in so he broke a window and gained access.

The policewoman was not on duty and she failed to report the crime. Why?

Clues

1. She knew the person's house.
2. She did not follow the incident up when she next went to work.
3. She liked the people living in the house.
4. She knew that the people living there were in no danger.

See answer 10

The Removal Men

The removal men had been asked to pack and move the contents of a very expensive house to another even more exclusive area. The house contents included fine silver and gold cutlery, rare pieces of art and very expensive collections of stamps. One of the removal men found the temptation too much and stole a page from the stamp collection. It was the homeowner who was jailed. How could this be?

Clues

1. It was not an insurance scam.
2. The removal man did not know the homeowner.
3. The removal man lost his job and was arrested.
4. The value of each of the stamps was over $10,000.

See answer 24

The Savage Attack

A man charged through a crowd of people and ripped off a pretty lady's blouse, punched her on the chest, and carried her away with him. The crowd were in shock and nobody tried to stop the man. Why not?

Clues

1. He had never seen the lady before.
2. The police pursued him.
3. He was not armed and was not a physically strong person.
4. The police did not arrest him.

See answer 53

Cheap Shopper!

A man on low income wanted more for his family than he could provide. He devised a scheme that he thought might help him achieve this. He was useful with a computer and understood how the supermarket system worked. After going to the supermarket he implemented his scheme. He had a full trolley of goods and was prepared to pay the price on the register for all of the goods, yet he was arrested. Why?

Clues

1. The register asked for $120.25, which he offered to pay.
2. All of the goods bought were in tins, jars, or packets. He did not buy any fruit or vegetables.
3. He had planned this very well and had not been noticed doing anything wrong by security cameras in the store.
4. He declared everything at the register and kept nothing in the trolley or on his person.

See answer 94

Father vs Son

Joe's son was very fit and worked out every day but he was not the brainiest of individuals. Joe had seen his youth come and go and he was now in his late 40s and not in good health. He felt that he could still beat his son even if he gave his son a small start. Joe's son, who would never throw a chance to beat his father, took up the challenge, but still lost. How?

Clues

1. Joe was never any good as an athlete.
2. Joe never cheated and did not have any help.
3. It did not involve any motors or sails.
4. Joe's son did not let his father win deliberately.
5. The son had a 10-second start.

See answer 72

The Awkward Piano

Alf was a bit of a practical joker and his workmates would always be under attack from him. One day they had to move a piano and some other items up the stairs in a department store. Although the piano was heavy, they decided that they could still put a few things on top of it before they carried it up the stairs. Alf was going backward and went up the stairs at the leading edge of the piano. Joe was at the bottom end and soon ran into a problem. Alf asked if he could hold the piano in place while he got help. Joe said, "Yes, but be quick." Alf rushed off and returned in under a minute pushing something into Joe's top pocket. "There," said Alf, "that should do it!" Joe was not amused. What had Alf done that he thought might have helped Joe so much? (Not!).

Clues
1. He used a literal translation of a need for help.
2. It did not help Joe at all and the piano was stuck.
3. Nobody else helped.

See answer 104

The Fire

The couple had just finished building their home and because the night would be very cold, they wanted to build a fire to keep warm. The wind outside was gusting at 40 mph and they were soon very cozy and fell asleep. A few hours later they were both dead. What had gone wrong?

Clues

1. The home had not burned down.
2. The house had not blown down.
3. They had not suffocated.
4. They had not been burned to death.

See answer 43

Is the Doctor Wrong?

A farm worker fell from his tractor and suffered bruising and what he thought might be a broken ankle. He was taken to the local hospital where the student doctor started to investigate his problems. Almost at once he shouted, "Cardiac arrest!" and revival equipment was rushed into the out-patients area. The diagnosis was correct and the farm worker went home in the next 5 hours. How could this be?

Clues

1. The farm worker was alive when he went home and he was discharged by qualified staff.
2. The student doctor did everything correctly.
3. The consultant physician thanked the doctor for his prompt action.
4. Neither the broken ankle, nor the bruising, caused the cardiac arrest.

See answer 113

The Master Forger

The best forger of all time was indeed a most brilliant artist and a man respected and sought after in the criminal world. He was so good that every major intelligence force kept a watchful eye on him and anyone he came in contact with. They even bugged his home and workplace with microphones and cameras. This came in useful when he was asked to copy the new $50 note. The police were tipped off and decided to search his premises before he even got started. Why?

Clues

1. It was not to see if he had any paper or ink.
2. It was not to see if he had any photographic equipment.
3. The search was successful.
4. The forger made perfect copies of the $50 note at a later date and was immediately arrested and jailed.

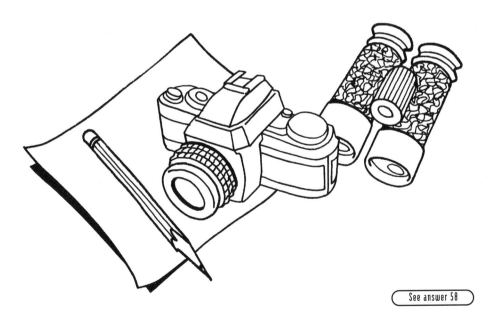

See answer 58

Golfers

Two golfers had a challenge match. One scored 72 and the other scored 74. The player with the highest score won. How could this be given the following clues?

Clues

1. They played off the same handicap.
2. They had both scored correctly.
3. Neither player had incurred penalty shots and they followed the rules precisely.
4. The player with the lower score was not disqualified.
5. It was not a tournament where only the player scoring 74 was entered.

See answer 87

The Cup of Coffee

A blind man went into a restaurant and ordered a cup of coffee. When it arrived he complained that the coffee was not hot enough and requested a fresh cup. When it arrived he complained that it was the same cup. How did he know?

Clues

1. The cup did not have a crack or anything that distinguished it from the other cups that were used in the restaurant.
2. He could not tell by the temperature of the cup.
3. He had not left a sticky mark or cream on the outside of the cup.

See answer 62

Mad Driver?

Why did the driver accelerate quickly to ram the car in front of him on the three-lane freeway?

Clues

1. He had not been drinking or taking drugs.
2. He did not know the driver in the car in front of him.
3. His foot had not had a muscle spasm; his action was deliberate.
4. He did not wish to harm anyone.
5. It was not a result of an act of nature such as an earthquake.
6. It was not to leap over a broken bridge or hole in the road.

See answer 31

Amazement

The child watched in total amazement as a man blew up a bank, killing three people. The child had a clear view of the whole event and was the only person to witness what went on. The police did not need to question him. Why?

Clues

1. The child was 12 years old.
2. The child told his parents what he had seen and they did not report it.
3. The family were not afraid of repercussions.
4. The man was not known to the child but he could describe the killer. and all of the events clearly.
5. The child was not one to lie.
6. The killer did not own up to the killing.

See answer 4

Triangles

What is the largest number of non-overlapping triangles that can be produced by drawing 7 straight lines?

This diagram only gives 5 but you can get many more from 7 lines.

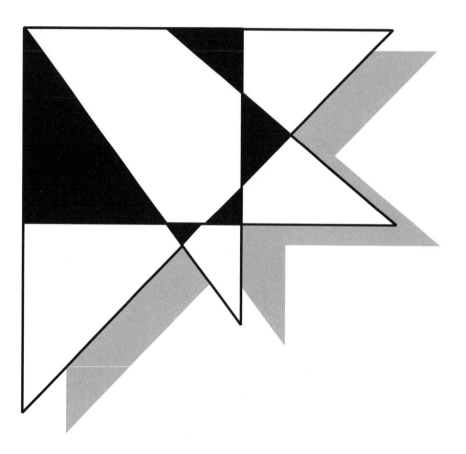

See answer 22

The Jealous Husbands

(This puzzle was devised in the 9th century by Alcuin of York)

Three jealous husbands with their wives have to cross a river in a rowboat. The boat can only carry 2 people at a time, and only 3 of the 6 people can row. How can the 6 of them cross the river so that none of the women will be left in the company of any other man exept when her husband is present?

See answer 59

The Meeting

The man from Nepal came by plane to visit the man from China who wore a chain around his neck. What was the weather like when the man from Iran joined them?

See answer 99

Confusion & Lies

There was once a family that was well known for being awkward. The males in the family always told the truth but the women in the family never made two consecutive true or untrue statements.

When met by a visitor, the father and mother had one child with them. The visitor asked the child, "Are you a boy?" but the visitor could not understand the reply. One of the parents said that the child responded, "I am a boy". The other parent then said, "The child lied, she is a girl". Was the child a boy or a girl and what did the child say?

See answer 78

Lateral Thinking Gem From Times Gone By

Can you make 101010 into 950 by adding just one straight line.

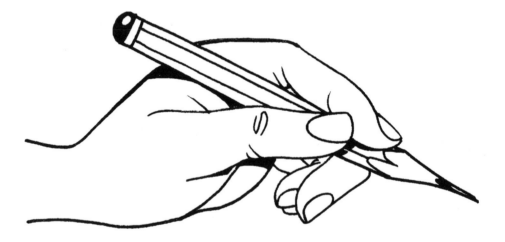

See answer 111

The Casino

Five people sat on the edge of a large casino and played from 10pm to 3am. They were professionals and did not stop for a break and nobody joined or left them. They played together without the assistance of anyone from the casino. They kept their own scores and, at the end, all of them went home with more than they had to start with. How could this be?

Clues

1. They were not playing against machines such as slot machines or blackjack machines.
2. They were not playing bingo or against the house.
3. Each of them went home not losing and always gaining. whenever they played together at the casino.

See answer 47

Corporal in the Army

A man sat down in a restaurant and started to read the menu out loud, but to himself. "Steak and fries, $7; steak, egg, and fries, $8.50; salad, $4 ..." etc. The waiter went up to the man and said, "You must be a corporal in the Army". He was correct but how did he make this connection?

Clues

1. They had not met before and the man was alone.
2. They were not near an army base.
3. The man's voice was not disturbing anyone.
4. He did not speak like a drill-sergeant.

See answer 112

The Abandoned

Charlie was abandoned at an early age and life had been a struggle, not just for him, but also for his adoptive parents. He killed his adoptive parents' offspring, yet they still worked hard to ensure he survived and had a home. As soon as Charlie was old enough he left his parents, never to return.

Neither the police nor the social services had anything to do with Charlie, even though he also killed his own offspring. Why?

Clues

1. It had nothing to do with being underage when he killed.
2. His family had a reputation for punctuality.
3. His adoptive parents did not press charges even though the murders were brutal.
4. He never joined the military or had a social service number.
5. He was born in the spring.

See answer 11

The Cheetah and the Hyena

The cheetah tells lies on Mondays, Tuesdays and Wednesdays, and tells the truth on each of the other days of the week. The hyena lies on Thursdays, Fridays and Saturdays, but tells the truth on each of the other days.

One day the lion heard them talking. The cheetah said, "Yesterday I lied all day," to which the hyena responded with exactly the same words. What day was it?

See answer 86

Lost For Days

What day is it when the day after tomorrow is yesterday and today will be as far from Sunday as today was from Sunday, when the day before yesterday was tomorrow?

See answer 65

Alien Conference

It was the year 2156AD and 1000 aliens attended the intergalactic meeting on Mars.

606 had 3 eyes.

700 had 2 noses.

497 had 4 legs.

20 had none of the above 3 traits.

4 times as many people had only 3 eyes as an oddity as had only 4 legs as an oddity.

220 aliens had a combination of all 3 oddities.

How many aliens had only 2 noses as an oddity if only 30 aliens had 3 eyes and 4 legs as oddities?

See answer 80

The Millionaire's Inheritance

A millionaire leaves $14,148,167 to his 7 sons and the rest to charity. In his will he makes a proviso that everything must be given to charity if the sons cannot divide the money equally between them. Is there a way in which they can inherit?

See answer 9

Another Mansion Murder

The Lord of the Manor has been murdered. The visitors to the manor were Abbie, Bobby and Colin. The murderer was the visitor who arrived at the manor later than at least one of the other two visitors. One of the visitors was a detective who arrived at the manor earlier than at least one of the other two visitors. The detective arrived at midnight. Neither Abbie nor Bobby arrived at the manor after midnight. The earlier arriver of Bobby and Colin was not the detective. The later arriver of Abbie and Colin was not the murderer. Who then was it who committed the murder?

See answer 21

In the Dirt

Two children were playing in the loft of a barn before it gave way and they fell to the ground below. When they dusted themselves off, the face of one was dirty while the other's was clean. Only the clean-faced boy went off to wash his face. Why?

Clues

1. Neither of them needed cold water to stop bruising and neither child was hurt.
2. Neither child put their dirty hands on their faces.
3. It was dusty and they had both sweated.
4. Their faces had not touched the ground.

See answer 52

The Holiday Disaster

Bill Drallam and his lifelong companion did not like the cold weather and often flew to the warmer southern states for a winter break. This year they decided that they would go with other friends in a larger group. They reached the airport and most of the group were killed, together with 30 people they had never met before. The survivors who suffered injury who were in their group were not taken to hospital, yet all of the other survivors with injuries were. Can you explain what occurred?

Clues

1. Members from their group caused the problem.
2. They did not cause the problem deliberately.
3. No disease or virus was involved.
4. It was not a terrorist or hijack situation.
5. It had nothing to do with guns.
6. If they had not gone with the larger group the 30 strangers may have survived.

See answer 91

Evolution

Three uninhabited islands were within swimming distance from each other but only at certain times of the year. This depended on the strong currents that flowed between them. A group of naturalist explorers put animal x on island A, animal y on island B and animal z on island C. No other animals were on the islands and no animals visited the islands.

When the explorers returned several years later they found island A had no animals on it. Island B had animals x and y plus one new animal on it, and island C had the same type of animals as island B plus z and another new animal. Can you name the five animals?

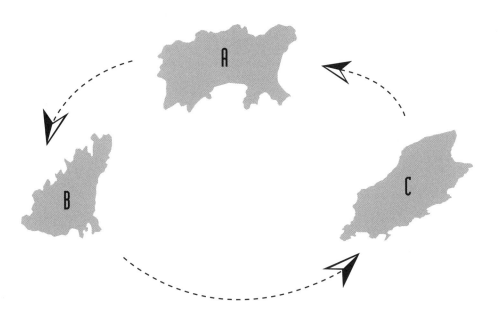

See answer 73

The Full Cask of Wine

Following a shipwreck a cask of wine is washed ashore and is lodged precariously on some rocks on the seashore. The sole inhabitant of the island only has a bottle with a rubber seal which fits the bunghole at the top of the cask exactly. He also has an endless supply of fresh drinking water. He cannot move the barrel at all and cannot break the cask for fear of losing all of the contents. How does he get the wine into the bottle if he is not allowed to put water into the cask and he does not wish to spoil the wine?

Hole at the top of the cask

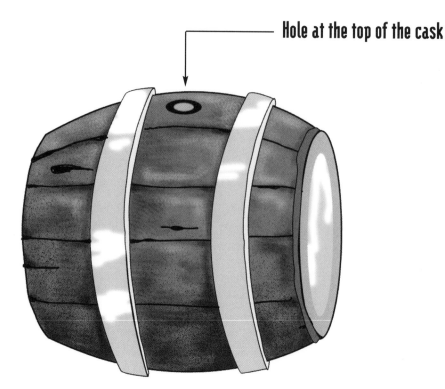

See answer 100

Recovering with a Letter

A deaf lady was tricked by a conman who told her that he could make her hear if she bought a special letter from him. When she opened the envelope what did she find?

See answer 46

The Twins Cause Confusion

A father always wanted 4 sons. His ancestors had always had large families and so he thought nothing about it. He was, however, upset in later life because he had only produced 3 sons. His eldest son was now 28 years old and he had given him a quarter of his land as his inheritance already. He had not passed other shares to his other sons before a wonderful event occurred; twins, and both boys! He immediately split the remaining land into four equally shaped parts, which were also equal in area, and gave each remaining child a share. How did he do this, given that he had divided the land awkwardly?

1st Son's Land

See answer 114

How to Trick the Genie?

The king had a magic lamp that contained a genie. He also had a beautiful daughter who loved Aladdin, but the king did not like Aladdin and did not wish them to marry. He did not wish to upset his daughter, so one day he rubbed on the lamp and devised a plan with the genie. The king said he would call upon Aladdin and his daughter and seek a test of worthiness from the genie for Aladdin. They would all have to abide by the results. Aladdin was passing by when he heard the king and the genie planning the event. The genie said, "I will produce two envelopes for Aladdin to choose his fate. We will tell him that one contains the words 'Get Married' and the other will contain the words 'Banished Forever.' Aladdin must choose one envelope, but I will make sure that both envelopes have 'Banished Forever.'"

How did Aladdin trick the genie and the king?

See answer 12

Car Grid

You are in a car that is parked and facing east on a straight road. You set off in the direction of the facing road and after some time driving you finish up 2.7 miles to the west of where you started.

How?

Clues

1. It is not a car with hovering capabilities.
2. It is not on a trailer or being towed.
3. You have not gone around the world.
4. You cannot turn the car around.

See answer 81

Does It Add Up?

Two mothers and two daughters went shopping for new dresses for a wedding celebration. They each returned with a new dress, but they had only bought 3 dresses. How can this be correct?

See answer 79

Not So Scientific

What is it that you can see with the naked eye, seems to have no weight and yet the more of them you put into an empty container, the lighter the container becomes? Two answers are possible.

See answer 30

The Arctic Explorers

A man went into a seafood restaurant and ordered seal stew. After only a few mouthfuls he wrote a note to the police and then pulled out a gun and shot himself. Why?

Clues

1. He was not an unhappy man and he had not contemplated this action before going to the restaurant.
2. He chose seal stew because he had had to survive on this food for 14 days on a recent expedition.
3. His note told the police that he was committing suicide and giving the reason.
4. The reason given referred to his last expedition with two friends.

See answer 15

The Gravel Quarry

Big Al and Little Joe had just robbed a jeweler's but the police were not far behind them. Their escape route went near an abandoned gravel quarry where Little Joe worked when the quarry was open. They stopped and dropped the bag containing the jewels over the edge of the rim and saw where it landed. Just to make sure it was well hidden, they threw some dry sand over the bag where it had landed. After 20 seconds they looked over the edge and they could not see the bag, and the sand blended with the damp sandy surface below. Two miles further on the police arrested the men and later had to release them for lack of evidence. Big Al killed Little Joe the next day and got away with it. What were the circumstances?

Clues

1. Neither of them had told the police where to find the jewels.
2. No animal, bird or person moved the jewels.
3. The jewels had gone from the spot where they were stacked.
4. Big Al did not take the jewels in the night and he did not suspect Little Joe of taking the jewels. Little Joe did not suspect Big Al of removing the jewels.
5. They remembered the correct spot exactly.
6. A warning sign had been placed so that it could not be seen from above.

See answer 60

The Silence

"Hello," said Henry, as he gave his girlfriend a peck on the cheek. He then asked, "Where's dinner?" After a few moments he said, "Your dad couldn't say that". He was right, but do you know why?

Clues

1. Her father was alive and in good health, and there was nothing wrong with his mind or voice.
2. He was born and raised in the same country as all of his offspring.
3. They still lived together and communicated every day.
4. The father was not angry with Henry.

See answer 97

Hit the Wrong Button

Adam was into all forms of sports and adventures. One day he joined an outdoor activity club, and after the necessary instructions had been given, he joined the rest of the group. The training was intense and his enthusiasm for the sport often clouded his judgment. Everything had gone well until he was on the return leg when he hit a button that caused his death. What was the sport and what button did he hit?

Clues

1. Training took a few weeks.
2. The sport required him to follow several safety rules.
3. Some people have been known to enter into the activity without training.
4. He was less than a mile from the finish when he made the mistake and was traveling at under 30 mph.

See answer 69

Little Annie

It was just after Christmas when Little Annie went to the village store to buy some candies and a few things for her mother. "That will be $10.50, Annie," said the storekeeper. Annie handed over a $10 bill and a $5 bill, and waited for her goods and change.

"I can't give you the goods or change until your mother comes in, Annie," the storekeeper explained in a very friendly tone. Why?

Clues

1. She had not purchased any tobacco, perfume, alcohol products, or anything where her age required her to be older.
2. Her mother would not have been upset with her even though she had not asked her to go to the store.
3. She often ran to the store for her mother for small items and was always partial to a few candies for her trouble.
4. She was an honest child.

See answer 102

The Magician

The magician's table is smoking with carbon dioxide gas, produced from dry ice in water. The mystery increases as he taps a smoking metal ball with his wand and places it in a wooden box, which is just big enough to enclose it. The box is placed on a tray for all to see and a few moments later the ball is gone. What was the scientific explanation for this?

Clues

1. It was a solid metal ball.
2. A small hole at the bottom of the box existed.
3. The ball was 30 times too big to go through the hole.
4. The box was hot.

See answer 14

Answers

1. Racetrack Confusion

It was his mother in car 3.

2. Uneasy Peace

The McPhersons were given the numbers : 5, 6, 7, 8, 9, 12, 16, 18, 19, 22, 23, 24, 26, 27 & 30. If the count started at number 1 all of the McPhersons had to jump overboard.

3. Sally's Wash

The pipe to the basin has frozen so the plughole was also frozen. Therefore, as long as Sally did not run the hot tap, the water would stay in the basin.

4. Amazement

The child saw it on a TV movie.

5. The Rejected Recruit

He was trained as a sharpshooter or sniper. His type of color-blindness allowed him to pick out other snipers wearing camouflage quite easily. He was therefore a very important member of his unit since he could see the enemy and get in the first shot. [This technique has been employed especially in jungle warfare.]

6. Nylon Ball-Bearings

He fell into a storage compartment and sank to the bottom. He eventually ran out of air.

7. The Disappearing Man

He had melted; he was a snowman.

8. Head-On Ant Crash?

They asked not to go on the rod at the same time.

9. The Millionaire's Inheritance

Only if they can express the number to base 9, which gives $7,000,000 or $1,000,000 each.

10. The Policewoman

It was her husband breaking into their own house after they had locked the keys inside.

11. The Abandoned

Charlie was a cuckoo. The punctuality clue refers to a cuckoo clock.

12. How to Trick the Genie?

Aladdin chose one envelope, and without opening it, tore it up into lots of pieces, and asked the King to read what option he had rejected in the other envelope.

13. The Musical End

The music and lighting were on the same circuit. The emergency and life-support systems were on another circuit. It was nighttime and when the music stopped the lights went off. In the confusion, one of the visitors accidentally disconnected some vital equipment and Grandpa died.

14. The Magician

The ball was made from frozen mercury, which melted and went through the hole in the base to a glass container. The box was left dry inside.

15. The Arctic Explorers

The three on the expedition were cut off by bad weather and had no emergency supplies. His best friend and his other colleague went for food. Only the colleague returned. For 14 days his colleague had told him that they had been eating seal stew. When he tasted seal stew in the restaurant, he realized that he had eaten his best friend.

16. Sinking Robots

The mass of the planet was much greater, although its size was the same. This meant that its gravitational forces were 10 times greater, the effect of which meant that the robots weighed 10 times what they would have done on ZOD. This caused them to sink to a level where they would not function.

17. Big Bill

Big Bill was a lighthouse keeper who had stayed awake the previous night to keep the light working in the worst part of the storm. The alarm bell on the buoy had been smashed on the rocks and no longer gave an audible warning. The lighthouse light was switched off in error and, as a result, a ship crashed on the rocks.

18. A Problem for the Ferryman

Nine trips are required. Label the children A, B, C, D, and E in ascending age, and the sides of the river "Near" and "Far" to create the table below:-

Trip No.	Near Side	Children in boat	Far Side
1.	A, C, E	B, D	None
2.	A, C, E	B	D
3.	B, E	A, C	D
4.	B, E	A, D	C
5.	B, D	A, E	C
6.	B, D	C, E	A
7.	B, D	C, E	A
8.	B, D	None	A, C, E
9.	None	B, D	A, C, E

Each child has had 3 one-way trips

19. Brother Simon

Brother Simon is a ghost and passes through the walls.

20. The Class

James is a teacher.

21. Another Mansion Murder

Abbie.

22. Triangles

23. The Courier's Wait

He was on a boat and had to wait for the next high tide to get into the unloading dock.

24. The Removal Men

The removal man took the stolen stamps to the biggest stamp dealer in the city, who recognized that they were stolen from his shop some years before. He called the police, who arrested both the homeowner and the removal man.

25. Hold-up Clues

The customer had hiccups. The fright of seeing the masked gunman did the same job as the water would have.

26. Little Breeders

They were all female birds.

27. The Immovable Screw

He was using a two-way screwdriver, which had a clutch action.

It had last been used to put a screw into a panel by turning the screw clockwise. The reversible switch to engage the screwdriver for undoing screws had not been altered.

28. Levitating Balloons?

They had left the bath running, which overflowed through the ceiling. The draught excluders prevented the water from escaping and the water level in the room was 2 inches deep.

29. Washing Dishes

The odds of the deciding lot would be the same for each round, and over time that child (unless unlucky) would be required to wash dishes on a Sunday as many times as each of the other children.

30. Not So Scientific

Holes or beams of light.

31. Mad Driver?

He saw someone cutting across from the opposite side of the road and they were spinning out of control and heading straight for him. He was boxed in, and rather than take a head-on impact, which might have killed both drivers, he took a minor bump on the car in front of him.

32. Target Practice

It had snowed overnight so they cleared the yard and made snowballs, which stuck to the barn and melted away afterwards.

33. The Burglar

The burglar had just burgled the house next door when he noticed that the neighboring house was on fire. He immediately entered the building to raise the alarm. Checking the rooms, he found two children overcome by smoke and took them to safety. The neighbors saw what had gone missing, and it was still in the burglar's hand when they called the police.

34. The Deadwood Stagecoach

He woke up (still drunk) on the stagecoach, which was still at the saloon, but it was the wrong one, and was not heading east.

35. The Fabric Shop

The curtains also have horizontal lines, so they are checked.

36. Antony & Cleopatra

They were both pet fish and the tank that housed them had a crack; all the water had leaked away.

37. Moving Suitcases

A violent tornado had ripped through their apartment and car-

ried the contents over a few miles. A kind lady found the cases in her yard; because the address labels had not been filled in, she decided to place them at the side of the road so that the owner might see them if they drove past. Because of the damage a few miles away, the police were keen to help the homeless and rescue services. That was the reason for not passing the property on to the police.

38. High Days and Holidays

Saturday, Wednesday, Thursday, Tuesday, Sunday, Monday & Friday (sum of alphabetical positions).

39. Disappearing Treat

The bag contained candyfloss. Rain got into the hole in the top of the bag, and the candy dissolved into a small amount of pink liquid.

40. A-Haunting We Will Go!

Take the room number at present multiplied by the number of days between the sightings, and then subtract the number of days between sightings. The number of days between sightings increases by one for each period. The next sighting will be $(9 \times 4) - 4 = 32$, thus Room 32 every fifth night.

41. Amateur Safe-crackers

Like the expression from the firing

of flintlock rifles "a flash in the pan," they had not compacted the powder or kept it in a container to cause an explosion. It therefore ignited like the powder in the "pan" of a flintlock rifle and just went up in a flash and a great cloud of smoke. In olden times the "charge" of powder was compacted inside the barrel of the rifle. A small charge of loose powder was placed in a small bowl where the flint's spark would ignite the powder in the pan, which then lit the compressed powder in the barrel through a small hole. Once ignited, the compacted powder would cause the bang and the shot to be fired.

42. Survival of the Weakest?

Your first shot should go behind you or deliberately in the air. You can't shoot at Count Nevermiss because if you did and were unlucky enough to hit him, Lord Bullseye would polish you off with the next shot or two. If you shoot at Lord Bullseye and hit, Count Nevermiss will certainly get you. If you miss Lord Bullseye, Count Nevermiss would not and his chances against you are 2 : 1 in his favour. If you hit Count Nevermiss, Lord Bullseye's probability of winning against you is 6/7, yours is 1/7. But if you deliberately miss, you will have another shot against either one of the other two. If Lord

Bullseye hits the Count, you will have a 3/7 probability. With 1/2 probability, the Lord will miss the Count (in which case the Count will dispose of the Lord). Thus your chances are 1/3 against the Count. The odds are increased by shooting in the air: the first shot will be 25/63 (about 40%). Lord Bullseye's odds become 8/21 (38%); Count Nevermiss's odds are 2/9 (22%).

43. The Fire

They were explorers who had built an igloo. The fire was too big and melted the walls when they fell asleep. They both suffered extreme hypothermia and died.

44. Happy New Year and Again and Again etc.

She was an astronaut, who on one occasion was in a stationary orbit over the International Date Line. As each date line revolved below her, she celebrated the New Year 26 times. The other times occurred while she flew from east to west, passing through three date lines when it was midnight on the ground.

45. The Strong Swimmer

A break in the seabed released large quantities of trapped air as small bubbles. This reduced the density of the water so that it was

lower than the density of a human body and he sank.

46. Recovering with a Letter

A piece of paper with the letter 'A' on it. The instructions said, "If you add 'A' to 'her,' you will have 'hear.'"

47. The Casino

They were a band who played background music for the guests.
They were paid by the casino and did not gamble.

48. Charged by a Bull

One of the family named BULL who owns the nearby farm charged them $20 to cross his land.

49. Don't Jump to Conclusions

He was a priest whose birth had been in the presence of his father.

50. Leap to Safety

His house is built into a hill/ the house is built below ground level.

51. The Aircraft

It was the earliest days of commercial jet aircraft flight and a few unexplained accidents involving the Comet needed to be investigated. The Comet was the first commercial transatlantic passenger jet. It flew higher and faster than all other commercial planes, and

was therefore subjected to stresses that other planes had not endured. The main problems came when the pressures in the fusilage were greater than those outside. The design engineers found that this was best simulated by putting water in the fusilage under pressure. This identified a number of weaknesses in the design, especially around the windows. The findings have made all jet travel much safer.

52. In the Dirt

One child fell on his feet, and his face was not covered with dust to make his face dirty. When he saw his friend's face covered in dust, he thought his own must also be dirty; his friend only saw his friend's clean face. The dirty child did not think that he needed to wash.

53. The Savage Attack

The lady was in a shopping mall and suffered a heart attack. Her heart had just stopped. The man who came to the woman's aid was a doctor just passing by. He started her heart, put her into his car, and drove straight to a nearby hospital with a police escort. The police were initially a little slow, and had to pursue him before clearing a way for him.

54. Who Is The Bigger Liar?

It was prisoner D. He lied twice but nobody said that he did not leave the cell for a few minutes to steal the sugar.

55. Trackside Jo

His bet was for a race two weeks earlier when the horse trailed in last. The newspaper gave the previous day's results when the horse had won.

56. The Tea Party

The little girl was in her playhouse. She had to go through the front door of the playhouse first, and then the front door of the family house to get to the front garden.

57. 1930s

It was a seaplane. The water that it was to use at the first landing sight was too rough for a safe landing so the pilot diverted to an airfield on land.

58. The Master Forger

They wanted to substitute the $50 note in his flat with a $50 note with a flaw in it. This flaw was unique to that note only, and when more of them hit the streets it could be traced back to him alone.

59. The Jealous Husbands

Men = ABC   Women = abc

Near Bank	Boat	Opposite Bank
ACac	Bb	None
Acac	B	b
ABC	ac	b
ABC	a	bc
Aa	BC	bc
Aa	Bb	Cc
ab	AB	Cc
ab	c	ABC
b	ac	ABC
b	B	Acac
None	Bb	Acac

60. The Gravel Quarry

The jewels were thrown on to quicksand. Little Joe had forgotten about it, but Big Al made him try to get the jewels. Little Joe tried and sank without trace. The police did not even know that Little Joe was dead.

61. Cardinal Lock'emup

One of the inkwells had disappearing ink. When the Captain of Guards saw nothing except the seals, he arrested the musketeers until he could speak to the Cardinal.

62. The Cup of Coffee

He had put sugar in the first cup.

63. The Messy Eater

He brings tinned fruit or pre-prepared fruit.

64. The Arabian Prince's Car

The clear plastic sucker that attached the air freshener to the front windscreen was shaped like a lens and focused the sun's rays like a magnifying glass on to the newspaper. The paper caught fire, causing the damage.

65. Lost for Days

Sunday.

66. The Puzzle King of Eygpt

The cube has two diagonal dovetail slots. The top can be pushed off by pushing at 45^0 to the face.

67. Lottery Winners

$230,000 ($15,000 increments)

68. Arise

It was an undersea mountain, and natural buoyancy lifted him.

69. Hit the Wrong Button

It was a parachute jump. He had hit the release button after his chute had been deployed.

70. The Great Soccer Player Retires

He played the whole of the first half, and for 10 minutes of the second half for his club. He scored 2 goals for his club in the first half. He was then taken off the field of play, and invited to play for his country for the last 35 minutes, scoring twice more. The other deciding goal was an "own goal" not scored by him.

71. A Bargain

The land was being reclaimed from the sea for industrial use. His company owned the reclamation contract. The land would soon be worth a fortune.

72. Father vs Son

Joe beat him at a game of chess (or something similar).

73. Evolution

Animal X on island A was an ass. Animal Y on island B was a horse. Animal Z on island C was a donkey. The new animal on island B was a MULE (ass/mare). The new animal on island C was a HINNEY (Donkey/Stallion).

74. The Last Train

The clock he saw was a reflection. It was showing 12:45, but this appeared to be 11:15 since the clock only had dashes on it rather than actual numbers.

75. The Bus Drivers

The two bus drivers are married; one is the boy's mother and the other is his father.

76. King-Elect

Only the less-bright child was a male.

77. The Bath of Liquid

He fell into a storage bath containing mercury. He was taken to hospital to be decontaminated because mercury can cause health problems. At room temperature mercury does not leave the skin feeling wet.

78. Confusion & Lies

If you say the child was a boy then the second speaker must have been the mother, whose first statement must have been a lie and whose second statement was true. But boys in the family do not lie so this option is no good. If you say that the child was a girl and if the first speaker was the father, then the second speaker was the mother whose first statement would be true and whose second statement was a lie. In that case the child would have spoken the truth and would have said, "I am a girl". But this implies that the first speaker lied, but males cannot lie. This option is therefore no good. So by deduction the first speaker was the mother and the child said, "I am a boy." The first statement from both the mother and child were lies. The child was a girl.

79. Does it Add Up?

They were grandmother, mother, and daughter. Two were mothers and two were daughters.

80. Alien Conference

107.

81. Car Grid

After going forward, you reverse.

82. The King Is In His All-Together!

It was a birthday parade where all of the participants carried pictures of themselves at birth.

83. Bush Fire

To put out the fire they used airplanes to scoop water out of the nearest lake. When they scooped the water out, they scooped him out as well. Water dropped on the fire and put it out but the fall killed the diver.

84. Household Enquiry

"Sally, are you in?" or "Are you there?" etc.

85. A Fruity Problem

The fruits are made of wax; they are candles, and the woman lit them before leaving the room so that they had burned down.

86. The Cheetah & the Hyena

Thursday.

87. Golfers

They were playing darts in the clubhouse. The object of the challenge was to see who could score the most with just 3 darts.

88. St. Joseph's Church

His father was the Italian Ambassador and he moved from Rome to Washington. Daniel only spoke Italian.

89. A Body Bag in the Suitcase

He was a part-time ventriloquist and it was his dummy.

90. My Homework is Right!

He was adding hours to his watch. 10 o'clock + 7 hrs = 5 o'clock.

91. The Holiday Disaster

Bill Drallam (Mallard backwards) was a duck. They flew in front of a plane during lift-off and entered the engine intake, causing the plane to crash. The plane might have survived if only one or two ducks flew into the engine, but several birds were hit and drawn into other engines.

92. Leaky Pipe

The second leak was halfway up the pipe. The first half was emptied in 1 hour, and with just the single leak left for the water to exit, it took another two hours.

93. The Inherited House

The sea had eroded the cliffs to within 30 yards of the garden. He found that it would be uneconomical to protect the house from further erosion. Experts had told him that it might only be 5 years before the mansion would be in the sea.

94. Cheap Shopper!

He had replaced all of the bar codes on the products with labels taken off small packs of the same items. The products he bought were all large packs and the bill should have been at least 3 times more. The shop assistant at the till raised the alarm when she saw one of the bar code labels was loose.

95. Two Brothers

The letter contained a white feather, a symbol of cowardice. In order to rid his family name of this slur, he was forced to act with bravery.

96. Dangerous Neighbors?

Mark. The policeman interpreted the question as "Question Mark Price!"

97. The Silence

Henry was the only parrot in the cage that could talk.

98. No Fire for Explorers

Neil and Dave were astronauts conducting an experiment on the Moon.

The lack of oxygen caused all of their problems.

99. The Meeting

Rain, for which Iran is an anagram (as Nepal is an anagram for plane, and China for chain).

100. The Full Cask of Wine

He washes some small pebbles and sand with the fresh water, and puts the washed and dried materials into the bottle. He then puts the bottle-neck into the bunghole. The pebbles and sand will fill into the cask to be replaced by wine into the bottle.

101. Problems With Air Pollution

Nobody lived to the east of the chemical plant.

102. Little Annie

Annie had been given a Monopoly game for Christmas and used money from its bank to purchase the goods. The storekeeper was not offended as he knew her very well.

103. The Share-Out

Child 1 had 10 25¢ coins, Child 2 had 16 10¢ coins, and Child 3 had 26 5¢ coins.

104. The Awkward Piano

The items stacked on the piano had fallen toward Joe, and he had said to Alf, "Give me a hand to move

them off!" Alf rushed into the department store and removed a hand from a mannequin and put it in Joe's top pocket.

105. The Unlucky Locksmith

The locksmith had been shut in with the manager after he had set the automatically activated system. He was inside the room just collecting the last items from the manager. The room had no light and the lock could not be tampered with from the inside.

106. The Fan

While he was celebrating it rained. The blue paint he had put on first was insoluble, but the yellow he put on top to create green was soluble, and had all washed off.

107. The Fire Drill

A swarm of killer bees had been sighted just outside the school.

108. The Bouquet of Flowers

1. 5
2. 17
3. 4
4. 10
5. 38

109. Car Park Overcrowding

Make all of the car park spaces at right angles to the wall.

110. Bob the Miser's Last Will

The Judge ruled that the money be shared equally between the relatives, but that they should each give Bob a money order for the money taken. If these were not cashed within 1 year of Bob's cremation, then the money could be kept.

111. Lateral Thinking Gem From Times gone by

10T010 (Ten to Ten, or 9:50 time).

112. Corporal in the Army

He was in uniform.

113. Is the Doctor Wrong?

It was the consultant physician supervising the student doctor who suffered the cardiac arrest. The student's prompt action saved his life. The farmworker was checked out, given a temporary plaster casing on his ankle, and was later allowed to go home.

114. The Twins Cause Confusion

115. The Glass Head

The two parts were lined up using strings with weights to guide the

head down. Several piles of sugar or other water-soluble materials were stacked at strategic locations on the plinth. The head was lowered and the ropes removed. The piles were then treated with a water spray starting from the central piles. (Dry ice could also be used.)

116. The Mountaineers

They were traveling to their destination by cruise ship. The hull of the ship was rammed during the night, and their cabin was below the water line. The pressure of the water held the door shut, they could not escape, and the rescuers were too late to save them.

Treasure Island

The rules of this treasure hunt are easy: all you have to do is find the intersection point of the open row and column.

Take the initial letters from the words in your answers to the following questions and delete that row or column.

1. I was given by France and symbolize truth and freedom in the USA.
2. I am the best-known structure in France.
3. The female monarch in England is one of these.
4. I am a famous puppet frog.
5. Goldfinger would love to live here.
6. Who owns this multi-colored coat?
7. Scientist famous for the Laws of Motion.
8. I did not say, "Play it again, Sam." But many think that I did.
9. Suffix to dates since Christ.
10. Margaret Thatcher held this top position.
11. 100.

See answer No. 23

The Swiss Deposit Code

A man carried a code for his Swiss account engraved in the buckle of his belt until he died. He did not pass on the secret to his family, but in his will he stated that whosoever cracked the code could have the contents of the safe deposit box in the Swiss bank. Can you crack the code?

DID	=	IIF
BAD	=	AG
EDFF	=	G+G
CE	=	A(B+E)
CCE	=	ACCB
BC	=	D+E-F
HEG	=	DBCC + GG
No		F - C - G - B
		OPENS THE BOX

See answer No. 87

Train the Train Driver

A rail depot had an oval track with two branch lines. This was used to train drivers in unusual conditions. The teacher gave them the following problem on the blackboard:

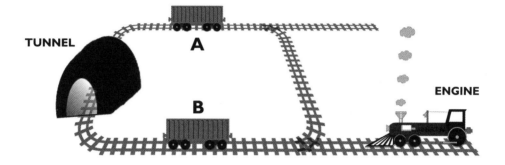

Move load A to position B and load B to position A without the load going through the tunnel, and return the engine to its starting position. How did the trainee drivers do this?

See answer No. 12

Take a Second Look

This is a series of letters in common use. Can you determine the next in the sequence?

N, W, H, O, I, I, ?

See answer No. 113

Square Meters?

If the perimeter of a rectangular field was 3000 meters, what would be the maximum area that you could contain within that perimeter if you could reorganize it into any configuration?

See answer No. 69

Fact or Fiction

The early Roman calendar originated in the city of Rome about 7 to 8 centuries before the Christian era. It was supposedly drawn up by Romulus, brother of Remus, in the February of his 21st year. Today modern historians dispute the validity of this. Do you know why?

See answer No. 32

Moving Water Uphill

You are given a dish of water, a beaker, a cork, a pin, and a match.
You have to get all the water into the beaker. You cannot lift the dish of water or tilt it in any way, and you cannot use any other implement to move the water into the beaker. How is this achieved?

water

See answer No. 55

Secret Messages

A journalist had been recruited by a foreign power to find out the chemicals being used in a top secret project. He was not given any contact name to pass the information on to. He was told to disguise the chemicals within a note in the personal column in the newspaper on April 1st and they would crack his code and obtain the knowledge. Only the journalist knew how he would transmit the message and the code he would use, but the foreign power knew that the message contained the name of 1 gas and 6 other elements or chemicals. The message was contained in the following text:

> *Jacob – Alter August's trip to Germany*
> *to the unfair one on the Nile.*

Can you find the hidden information?

See answer No. 128

Mysteries of Time

A young man proclaims, "The day before yesterday I was 17, but I will be 19 this year." Is this possible?

See answer No. 96

Logical Thinking with Matchsticks

By removing 4 matchsticks can you rearrange those left so that the top and bottom lines still have 9 matches in them?

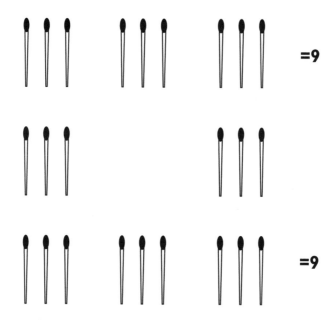

Only the most devious of lateral thinkers will find a second way to do this. Can you?

See answer No. 3

More Matchstick Trickery!

a) By moving just 2 matchsticks, can you increase the number of squares by 2?

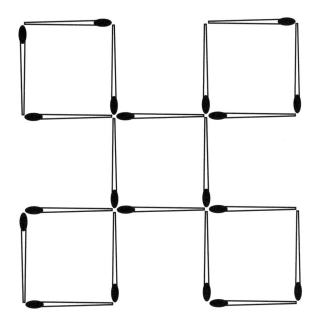

b) By moving 1 more matchstick, can you increase the square count by another 2?

See answer No. 74

Front Foot Forward

A man's right foot was facing due north and his left foot after one pace was pointing south. How was this possible?

Clues

1. The one pace was taken in the direction of the right foot and he did not turn in mid-stride.
2. His feet both pointed in the same direction.
3. His right foot had not been twisted around when it had been initially planted on the ground.

See answer No. 40

Triangles

The sequence of numbers 1, 6, 15, 20, ... continues with which three numbers?

See answer No. 103

A Waiter's Lot Is Not a Happy One

A waiter is serving vegetables to 51 diners in a hotel. There are peas, carrots, and cauliflower. Two more diners want peas and carrots only than those who wanted just peas. Twice as many people want peas only as cauliflower only. 25 diners do not want cauliflower, 18 diners do not want carrots, and 13 diners do not want peas. Six diners want cauliflower and peas but no carrots.

a) How many diners want all three vegetables?
b) How many diners want cauliflower only?
c) How many diners want two of the three vegetables?
d) How many diners want carrots only?
e) How many diners want peas only?

See answer No. 35

The Train Driver

You are driving a train. It stops at Milton Keynes and 25 people board it. It then goes to Leicester where 55 people get on and 43 get off. The next stop is Nottingham where 3 people get off and only 1 gets on. The train continues its journey, making Doncaster its next stop, where 19 get on and 13 get off. The next stop is York, which is the final destination. The driver then gets off the train also, and looks in the mirror in the washrooms. What color eyes does the driver have?

No clues for this one. It should be easy.

See answer No. 122

Logical Deductions of Who or What Am I?

1. What am I?

   Sometimes I am one before I'm one.

   When I'm under one I, and others of my species, are given the same name.

   Males and Females of my species have different titles.

   When I am over one but remain young these names change.

   Between the ages of one and two males and females of my species can also be given the same name.

   When I am fully grown I am called another name.

   All through my life people give me a name that is personal to me.

   I am eight and male. How am I known?

2. Who am I?

   I am deceased, but my name and actions are well known.

   I was a leader of my people but I never had a crown.

   I was often upset with my people and they were sometimes upset by me.

   I passed on messages and rules.

   I warned people of death and destruction.

   My most famous work was in stone.

See answer No. 101

Logical Deductions of Who or What Am I?/continued

3. **What am I?**
    I have been around for over a thousand years, but my appearance and format has changed with time.
    I have been mechanical in construction since the beginning and have taken electromechanical form since the 1930s.
    I have been miniaturized in my current form and I am used by almost all schoolchildren and adults alike.
    You have counted on me to help you for years.

4. **What am I?**
    I give birth, but I am the male of the species.
    I am covered by consecutive rings of body armor.
    I have a long tubular snout and live in warm waters.
    My eyes can work independently of one another.
    I am not a mammal.

5. **What am I?**
    I am a ballroom dance.
    I was made popular in the 1940s in Western Europe and USA.
    I have simple forward and backward steps with tilting and rocking body movements.
    I am danced to in 4/4 time with syncopated rhythm.
    My point of origin in S. America would give my name away too easily.
    I am a happy dance that is very popular in my country of origin.

See answer No. 101

Cleaning Confusion

At a dry-cleaner's one more customer brings in a jacket only than trousers only. Three times as many people bring in trousers, jacket, and a skirt as bring in a skirt only. One more person brings in a jacket and a skirt but no trousers than a skirt and trousers but no jacket. Nine people bring in trousers only. The same number of customers bring in a jacket only as trousers and a skirt but no jacket. 32 customers do not bring in a skirt and 24 customers do not bring in a jacket.

a) How many customers bring in all 3 of the items?
b) How many customers bring in only 1 of the 3 items?
c) How many customers bring in a jacket only?
d) How many customers bring in 2 of the 3 items?
e) What is the total number of customers bringing in any of the 3 items?

See answer No. 80

A Lewis Carroll Gem

The Governor of Kgovjni wished to give a very small dinner party, and invited his father's brother-in-law, his brother's father-in-law, his father-in-law's brother, and his brother-in-law's father. What is the minimum number of guests invited?

See answer No. 14

Word Connections

In each of the following sentences there are associated words that are hidden. These words can be found by looking at the end of one word somewhere in the sentence and connecting it to the beginning of the next word. What are the connected words?

1. The ballot usually takes no more than half an hour, then the fun begins. The hi-fi attachments are in place for David's party. In the fridge there is a large samosa above the sandwiches and a chocolate gateau diagonally placed on the shelf below to give more space.

2. If anything is wrong with the replica shirts then management must be informed so that they can combat escalating problems.

3. While Grandfather sang an impromptu ballad, the wood on the campfire began to char perfectly and Grandmother joyfully recited age-old rumors.

4. According to forces protocol, Lieutenant Barnabas settles the cashbox error and ensures the young soldiers faces will be agleam.

5. The cream beret that I managed to drop all through the mud at the Mexico rally has been dry-cleaned and should now appear lovely and clean.

See answer No. 9

Confusing Paper Model

Using a rectangular piece of paper can you make the model shown? You can make 3 straight-line cuts to the paper, and the paper model must not be glued or held together with clips.

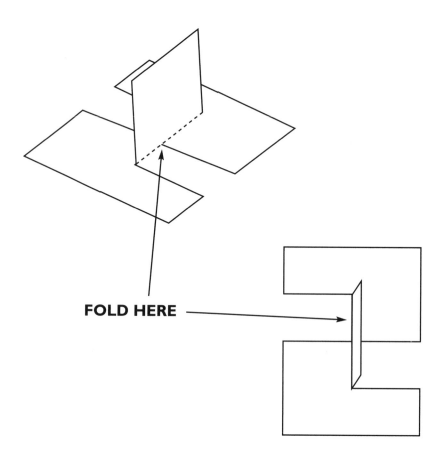

FOLD HERE

See answer No. 77

Confusing Family Relations

1. Alice, who had returned from Australia, wished to meet all of her relations, so she organised a reunion. She invited her mother, her mother's sister-in-law, her sister, her sister's mother-in-law, her mother-in-law's sister, her sister-in-law's mother and her next door neighbour. What is the smallest number of people that could attend the party if all invitations were accepted. No illegal relationships permitted.

2. A family has 5 children, of which half are boys. How can this be?

See answer No. 52

Look for the Simple Solution

What is the final product of this series of multiplications if all of the letters have the same value as the number of their position in the alphabet?

(t - a)(t - b) (t - c) .... (t - z)

See answer No. 19

The Mississippi Gambler

The professional gambler only played dice but he had made his own. He had 3 colored dice, and each color had 3 numbers, which were each on 2 faces.

The Red Die :   2 - 4 - 9 - 2 - 4 - 9   (total 30)
The Blue Die :   3 - 5 - 7 - 3 - 5 - 7   (total 30)
The Yellow Die :   1 - 6 - 8 - 1 - 6 - 8   (total 30)

The dice had not been loaded with weights. The gambler always let his customer have the choice of color, then he would choose his. It was always a game for only 2 players and the object was to have the highest number on any side.

How did this work so well for him? He always seemed to have an edge. Can you work it out so that by the law of averages he always had a better than 50 : 50 chance, and can you state what the real chances of his winning were?

See answer No. 93

The Fairground Game

At a fairground game, players had 3 darts each to attempt to win a teddy bear, a board game, or a beer glass. There were 4 winners of teddy bears and games not but glasses. Two more people won both a glass and a teddy bear but no game than those who won a glass and a game but no teddy bear. 43 of the prize-winners did not win a teddy bear, and 48 of the prize-winners did not win a game. Nine people won both a glass and a teddy but no game, and 31 people did not win a glass. 74 people won at least one prize.

    a) How many people won a teddy bear only?
    b) How many people won all 3 prizes?
    c) How many people won a glass only?
    d) How many people won 2 prizes only?
    e) How many people won a game only?

See answer No. 117

So You Think You're Good at Math?

Can you rearrange the following addition to make an answer of 100? You can use each number only once but you can add any mathematical symbols you wish.

$$\frac{6\ 1}{1\ 8}$$

See answer No. 8

The Striptease Artist

The man called the striptease bar from his home and asked for a particular entertainer to go to his room for a private session. After one hour the man said that he really enjoyed what she did and he felt much better. The man was the manager of the bar. He paid her to visit him and as a result he went to work.

Clues
1. She was only required to remove a few clothes.
2. The entertainer was between 18 and 20 years old.
3. She was not there for her beauty.

See answer No. 63

The Car Problem

When you are moving forward in your car are there parts of the car that appear to be going backward while being attached to the car?

See answer No. 125

Dozy Policemen

A small boy was riding his bicycle around the housing estate where he lived. He went up and down roads that had no outlets, in and out of trees, up and down the curbs. Unfortunately, he took a sharp turn and the front wheel of his bicycle hit a curb. The small boy fell from his bicycle and was knocked unconscious. Fortunately, there was a policeman at the scene of the accident, You would have thought that an ambulance might have been called, and all details of the accident and statements from anyone who witnessed the accident taken. Why was this not done?

See answer No. 81

Decimated

In Roman times, soldiers who were to be punished were forced to form a line, and every tenth one was executed. This is the origin of the word "decimate". If you were one of 1000 soldiers lined up in a circle, with every second soldier being executed until only one remained, in which position would you want to be in order to survive?

See answer No. 70

Children's Age

A man has 9 children born at regular intervals. The sum of the squares of their ages is equal to the square of his own. What are the ages of the children?

See answer No. 106

Random Chance

You have been blindfolded and asked to put a red sock in to the red bag, a blue sock into the blue bag, a white sock into the white bag, and a yellow sock in the yellow bag. The 4 bags and 4 socks are correctly colored. What are the odds that you can get only 3 of the 4 matched first time?

See answer No. 27

That & This

Add 'This' to 'That', divide by 3.
The square of 'This', you'll surely see.
But 'That' to 'This' is 8 to 1.
So figure what they are for fun.

See answer No. 33

The Warehouse Sale

I went to a warehouse sale and bought 3 lots of tee-shirts. The total cost was $260. Each lot was a different price and a different size. In each lot the individual price of the tee-shirts in cents was the same as the number of tee-shirts in the lot. If I bought 260 tee-shirts can you tell me the lot sizes?

See answer No. 98

The Rector Total

If each letter is substituted for a number in this addition, can you determine what the value of each letter should be?

```
        CELLAR
        CORPSE
        COLLAR
         CLOSE
          CASE
    +     COPS
        RECTOR
```

See answer No. 1

Cocktail Sticks

1. By moving only 3 of the cocktail sticks in the shape below, can you make 4 equal triangles? All of the cocktail sticks must be used.

2. By moving only 3 of the cocktail sticks in the shape below, can you make 7 triangles and 3 diamond shapes?

3. Using 6 matchsticks of equal length, create a shape with 4 equilateral triangles.

See answer No. 43

Cocktail Sticks/continued

4. By moving only 2 cocktail sticks, can you rearrange the shape below so that you will be left with 8 squares of the original size?

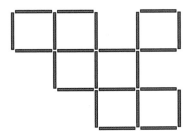

5. By moving 2 cocktail sticks in this arrangement, can you form 15 squares?

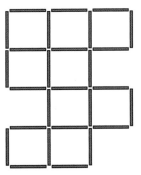

See answer No. 43

Rotations

Using the logic of the first grids, complete both of the incomplete grids.

a)

10	18	3
7		9
2	4	11

	4	
11		18
	3	

b)

T	D	M
L		E
K	U	Z

	U	
E		T
	Z	

See answer No. 53

Complex Numbers & Letter Grids

Can you find the missing numbers or letters from the grids below? The rules used in the completed grids, in each case, will give you the rules for the incomplete grids.

1.

	A	B	C	D	E	F
a	7	9	6	5	3	3
b	4	6	3	7	0	3
c	9	2	4	1	1	4
d	5	8	2	7	2	6

7	7	5	6	1	9
4	9	6	6	0	0
3	5	1	9	0	6
8	9	4	6	?	?

2.

	A	B	C	D
a	F	D	N	V
b	J	I	O	Z
c	M	Q	H	Q
d	G	A	L	Y

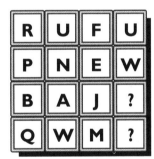

R	U	F	U
P	N	E	W
B	A	J	?
Q	W	M	?

See answer No. 38

Complex Numbers & Letter Grids/continued

3.

	A	B	C	D	E	F
a	7	8	3	5	7	9
b	3	7	4	5	2	9
c	2	2	1	2	2	2
d	4	2	7	5	0	8
e	6	5	9	8	6	4
f	8	2	1	7	5	6

9	1	6	8	4	5
8	3	2	8	8	2
3	0	?	3	1	1
0	9	?	4	9	9
6	4	9	9	1	5
7	1	4	9	6	7

4.

	A	B	C	D
a	5	9	10	16
b	8	8	5	4
c	4	36	2	8
d	10	2	25	8

20	14	8	12
3	4	16	4
6	7	?	6
10	8	4	8

See answer No. 38

Changing Words

Changing only 1 letter and making a new word each time, can you find the shortest routes between the 2 given words to change the first word into the second word? The order of letters must not change.

1.	**SEAT** – **TRAM**	(3 letter changes)
2.	**HEAD** – **TAIL**	(4 letter changes)
3.	**STONE** – **BRICK**	(7 letter changes)
4.	**WHITE** – **BLACK**	(7 letter changes)
5.	**HERE** – **JUNK**	(5 letter changes)
6.	**FAIR** – **RIDE**	(6 letter changes)
7.	**WRITE** – **CARDS**	(5 letter changes)
8.	**BROWN** – **TREES**	(4 letter changes)
9.	**GLASS** – **CHINA**	(6 letter changes)
10.	**GREEN** – **BLACK**	(6 letter changes)

See answer No. 84

Magic Squares

Can you complete these 2 magic squares so that each of the following items total 34? You must use each of the numbers 1–16 once only.

The rows across	**= 34**
The columns down	**= 34**
The cross diagonals	**= 34**
The center 4 numbers	**= 34**
Each corner block of 4 numbers	**= 34**

1)

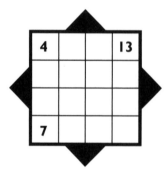

2)

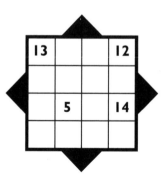

See answer No. 21

Extinct? I Don't Think So

Can you think of an animal, which, if made extinct and all of the seeds from that animal were also destroyed, could repopulate the world in under 2 years?

See answer No. 92

The Fire Station Location

The drawing below represents the time it takes to go between towns for the fire engine. You have to locate a fire station that minimizes the travel times to each location. Where would you locate the station to minimize the longest journey?

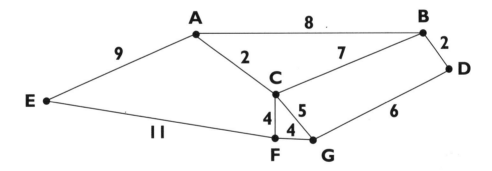

See answer No. 48

The Rabbit Family

How many male and female rabbits are there in a family if each male has 1 fewer female relative than he has male relatives, and each female has 2 males fewer than twice the number of female relatives she has?

See answer No. 26

Strange But True!

Prior to the American Civil War two famous people challenged each other to a duel. When the seconds had been selected the weapons were chosen. Pistols were suggested, but one of them objected saying that this was most unfair to him. One of the duelists was much taller and was therefore a larger target, whereas the other was a smaller target. How was this resolved?

Clues
1. The suggestion came from the shorter man and his seconds.
2. They could still fire at the same time and in the traditional way.
3. They were both given only one shot.

See answer No. 37

The Gearbox

The gearbox below consists of 4 gearwheels with intermeshing gears and two pulley belts, or drive belts. If the large 48 tooth gearwheel rotates exactly 10 times in a clockwise direction, in what direction will the pointer be facing on the gearwheel at the bottom of the arrangement, and how many times will it rotate?

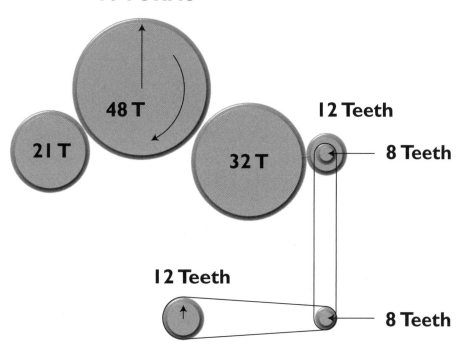

10 TURNS

48 T

21 T

32 T

12 Teeth

8 Teeth

12 Teeth

8 Teeth

See answer No. 123

Links?

What numbers should replace the question marks?

1.

8752	4524	1080
6978	5382	4346
7388	6424	?

2.

7628	5126	3020
9387	6243	1088
8553	2254	?

3.

9337	56	-1
8725	62	4
4821	?	-5

See answer No. 99

Equal Segments

Divide each of these 2 grids into 4 identical shapes so that the sum of the numbers in each section is as given below.

1)

TOTAL 50

8	8	3	6	5	5
8	4	4	7	7	4
5	5	5	8	3	5
9	8	3	4	7	3
7	5	9	3	5	8
6	4	4	8	3	4

2)

TOTAL 60

9	6	8	8	2	5
7	4	5	8	6	9
8	8	7	7	6	9
9	6	8	9	6	8
6	4	6	8	5	4
7	7	5	8	7	5

See answer No. 31

Classic Car

An Englishman who was a collector of classic cars was invited to display one of them at a rally in the USA. He noticed that he was not getting the mileage per gallon of fuel in America that he got in England. Why?

Clues

1. The fuel was the same octane from the same oil company.
2. He had tested it on long and short distances.
3. It was not due to hills.
4. It had nothing to do with humidity.

See answer No. 78

Time to Settle Up

What is the smallest number of checks needed to settle the following debts, if all debts are paid by check?

Ann owes Penny $20. Penny owes Mary $40. Mary owes Clair $60 and Clair owes Ann $80.

See answer No. 62

Numerical Links

The 2 words in the puzzle below have a connection with the numbers. Can you identify the missing numbers?

C	73	H
H	289	O
U	882	U
R	685	S
C	34	E
H	?	S

See answer No. 5

The Eleven-Card Con

The object of the game is to lift the last card. Two players can take either 1 or 2 adjoining cards each time. Player 1 starts from the center. If Player 1 removes 1 card, then you remove 2 cards from the opposite side of the circle of cards so that the cards are split into 2 groups of 4 cards each. If Player 1 removes 2 cards on the first move, you only remove 1 card but split the cards into 2 groups of 4. What strategy does the second player now need to adopt to win virtually every game? Remember, 2 cards can only be lifted if they are next to each other.

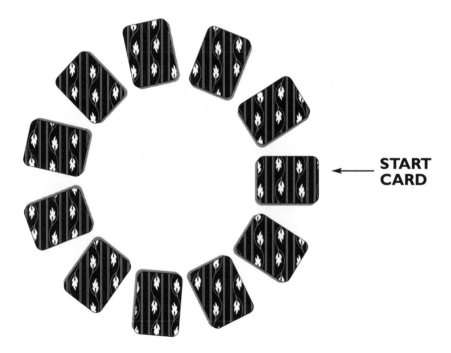

START
CARD

See answer No. 88

Blanket

A man walked to the top of a hill carrying a blanket. Over 100 people died because of this. Why?

Clues
1. People did not die of suffocation.
2. The blanket was clean to start with but became very dirty.
3. Neither the blanket nor the man carried any diseases.
4. It was warm work.

See answer No. 45

The Rail Workers

Two rail workers were repairing a line when an express train came thundering toward them at speed. The train driver was not aware that they were working on the line so he did not have the chance to slow down. The two workers ran straight toward the express train between the lines that the express was using. Why?

Clues
1. They had not gone mad and did not wish to die.
2. They had forgotten the express was due.
3. They would have been killed if they ran the other way.

See answer No. 102

The Detective Booking Clerk

The policeman and his wife went to a ski resort in Colorado. The policeman's wife was found dead at the foot of a large drop at the edge of a cliff. The booking clerk who had organized the holiday contacted the local police and the husband was arrested for murder. How did the clerk know it was foul play?

Clues
1. The clerk had never met the policeman or his wife.
2. The local police would not have arrested the policeman without the clerk's information.
3. The ski tracks did not show any foul play.
4. She died from the fall.
5. She was good on skis.

See answer No. 39

The Books

A lady walked up to the counter with 2 books and the assistant said, "That will be $6.95, please." The woman handed over the money and walked away without the books. Why was that?

See answer No. 15

Skyscraper

A window cleaner was cleaning windows on the fifteenth floor of a skyscraper when he slipped and fell. He suffered only minor bruising. How was this possible when he did not have his safety harness on and nothing slowed his fall?

See answer No. 116

Not Really a Premonition

A farmer went into a field and into a very large barn-like structure on the edge of the field. Later he came out, and as he was approaching the middle of the field he knew that when he reached it he would be dead. How did he know that?

Clues

1. It was not the farmer's field.
2. He had been to the same field on numerous occasions.
3. He did not do any work on the field.

See answer No. 30

Friend

Mary walked down the street when she met an old friend from her school days.

"Hello, I have not seen you since graduation in 1980," said Mary. "How are you nowadays?"
"Well, I got married in 1990, and this is our son."
"Hello," said Mary, "and what is your name?"
"It's the same as Daddy's," said the boy.
"Ah, so it's Robert, is it?" said Mary.

How did she know when her friend had not said the boy's name?

See answer No. 10

On the Farm

A group of children visited a farm and saw:

a) One animal with half as many letters as its plural
b) One animal with half as many letters as its young
c) One animal with the same sound as its plural

When the group moved further on they saw the same again but each animal was different. Can you work out what the animals were?

See answer No. 50

Transmogrification

An insect makes a beeline for, and attaches itself to, a fruit. What was the fruit if the combination of the two has become a widely used tool for builders and decorators?

See answer No. 72

Poisonous Insect

A lady saw a poisonous insect crawl into a hole in the wall behind her television set. Fearing for the safety of her children she wished to rid her family of this danger to them. It was late in the evening and she did not have any chemicals that would kill the bug since she hated to kill any living being. She did not wish to damage the house by digging into the wall. How did she get the bug without killing it?

See answer No. 108

Miss Punctual

Miss Punctual was, as her name suggested, always on time. She would always get up in the mornings at the same time, get herself prepared for work, leave the house on time, get to work on time, in fact she worked to the clock and you could tell the time by what she was doing. One morning, however, she was awakened by her alarm clock buzzer and she felt that her biological clock was out of phase with the world. She got herself ready and went to work, but arrived over 30 minutes late. She apologized, and said that she now felt well. What went wrong?

Clues

1. She was not ill.
2. She had not set the clock differently.
3. As soon as she got to work she felt fine.
4. Each of the things she had to do to get from bed to work took the same time as normal.

See answer No. 13

Holes In Paper

If you fold a piece of paper in half and cut a hole along the fold you will have one hole in the paper when it is unfolded. If you fold the paper in half, then half again at right angles and repeat so that you have made 6 folds, and then cut a hole in the last folded side, how many holes will you have in the paper when you unfold it? Try it in your head before reaching for the scissors.

See answer No. 28

Washers

How can you, by making only one single measurement, calculate the shaded area of a circular washer?

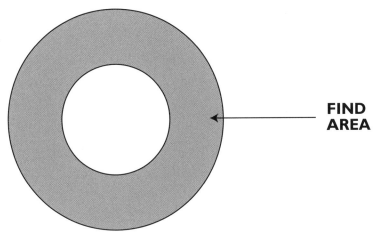

FIND
AREA

See answer No. 2

Hidden Words

Read the sentence carefully to discover a series of associated words hidden in it. The connected words will be found by joining the end of one word somewhere in the sentence with the beginning of the next word. What are the connected words?

1   A large wooden spoon to stir is so important, in the non-stick saucepan syrup will scorch identically to other sugary substances and will taste as terrible as you can imagine.

2   For those in even the slightest doubt whether honesty is the best policy, remember the trouble near the beacon goalposts by the gang escalates into something we are unable to control.

3   It took much effort of mine radically jumping into the cold river to rejoin eradicating the myth that I could not do it.

4   Please will all staff leave any extra bags in front of the coach as luggage was placed behind it and forgotten last time, and this was not realized until we arrived at Oslo customs.

5   At speed the little cherub ran off landing headfirst in the bathtub under all the soapy water but jumped out completely unaffected to astound everybody.

See answer No. 94

Hidden Connections

In each case, the following words have a hidden connection. What is it?

1 BEAUTIES SHOESTRING WHATEVER BELTING

2 DECODING ZEROES SPIKED SHAKER

3 DISPLAYED DRAMATICALLY FULFILMENT SHOWERING

4 SHAPELY ACCELERATE COMMANDEER WHELKS

5 MALARKEY CROOKED GROWLING TRAILERS

See answer No. 71

More Logical Deduction

What number should replace the question mark?

1	Baltic	= 151		2	Dill	= 601
	Arctic	= 201			Cumin	= 1101
	Ionian	= 2			Cardamon	= 1600
	Caspian	= ?			Aniseed	= ?

3	California	= 1524		4	Lemonade	= 1550
	Connecticut	= 301			Coffee	= 100
	Maryland	= 1550			Milk	= 105
	Texas	= ?			Cola	= ?

5	Chocolate	= 250
	Biscuit	= 102
	Cake	= 100
	Muffin	= ?

Look carefully at the letters in each of the words and work out the connection with the values given. What number should replace the question mark ?

6	Tagliatelli	= 40		7	Soccer	= 6
	Pizza	= 16			Baseball	= 9
	Macaroni	= 32			Golf	= 3
	Spaghetti	= ?			Tennis	= ?

See answer No. 25

More Logical Deduction/continued

8   Typewriter = 48
    Square   = 36
    Triangle   = 36
    Sphere   = ?

9   Circle   = 42
    Telephone = 84
    Printer   = 42
    Computer   = ?

10 Blackcurrant = 108
    Strawberry   = 72
    Gooseberry   = 144
    Pineapple   = ?

11 Soap   = 14
    Shampoo   = 28
    Toothpaste = 42
    Deodorant = ?

12 Watch   = 44
    Ring   = 33
    Necklace = 55
    Bracelet = ?

13 Doctor   = 108
    Nurse   = 81
    Architect = 162
    Lecturer = ?

14 Piano   = 28
    Clarinet   = 70
    Harpsichord = 112
    Guitar   = ?

15 Turquoise = 140
    Green   = 105
    Tangerine = 175
    Brown   = ?

16 Toe   = 4
    Neck   = 7
    Shoulder = 13
    Leg   = ?

17 Coat   = 12
    Hat   = 7
    Scarf = 9
    Gloves = ?

See answer No. 25

More Logical Deduction/continued

18 Perfume = 36
Mirror = 32
Brush = 28
Comb = ?

19 Pencil = 22
Paper = 18
Pen = 11
Staple = ?

20 Basin = 25
Bath = 23
Shower = 32
Toilet = ?

21 GUARD and ARTIST = 138
TAILOR and VICAR = 128
PILOT and MASON = ?

22 RUMBA and SAMBA = 19
TWIST and TANGO = 34
WALTZ and BOLERO = ?

23 ALMOND and WALNUT = 5369
BREAD and TOAST = 2250
YOGURT and CREAM = ?

24 FIRE and HEARTH = 196
CHAIR and SOFA = 160
BED and PILLOW = ?

25 CRICKET and BAT = 46
TENNIS and RACQUET = 83
SOCCER and BALL = ?

26 BRANDY and GIN = 68
SHERRY and WINE = 84
SHANDY and BEER = ?

27 MALTA and IBIZA = 70.5
CORFU and MILO = 87.5
TAHITI and HAWAII = ?

28 WYOMING and VERMONT = 320
GERONA and TOLEDO = 202
LOIRET and ARTOIS = ?

29 OVEN and STOVE = 3217
FORK and SPOON = 2579
PLATE and DISH = ?

30 FROST and SNOW = 1043
MIST and FOG = 2937
SLEET and RAIN = ?

See answer No. 25

178

Devious Clues

1   Two couples were in a jeweler's choosing a ring. If Derek and Janice chose a jade ring, what type of ring did Alan and Ophelia choose?

2   Kathy works in a BANK, and enjoys OPERA and SPORT. Katy is TWENTY years of age. Katy's partner's hobbies are ROWING and DRAMA, he is a BAKER by trade, and is THIRTY years old. What is his name?

3   Kim is hoping to receive some cards today. She has sent Valentine cards to Jack, Luigi and Tom. Who has sent cards to Joshua, Julian, and Sean?

4   Two holiday show presenters were sent to report on various locations. Diana visited Denmark, Nigeria, Spain, France, and India. Peter reported on Peru, New Zealand, Antigua, and Luxembourg. Was his final assignment in America or Australia?

5   The lady selling shoes is called Sarah. She lives in Chester but works in Bootle. She has a pet Doberman and plays squash. Fiona lives in Ormskirk but works in Southport. She has a pet Pekinese and goes yachting. What does she sell?

6   Ingrid enjoys SWIMMING, she has a pet CANARY, and her lucky number is SEVEN. Was Ingrid born in November or December?

7   Carol likes lemons but not bananas, opals but not emeralds. She likes red but not green, and she likes asters but not pansies. Does Carol like tea or coffee?

See answer No. 95

Find the Link

1   Four of the words in the diagram are linked by a common factor.
    What is it?

TRAP	–	**SHOUT**
FRIEND	–	**LIVED**
BOOK	–	**REGAL**
BATS	–	**PIG**

2   Does the word WOLF belong on the left or right side of this grid?

BEST	–	**PLEA**
DENS	–	**SODA**
NOSY	–	**LIFE**
FIST	–	**POLE**

3   Sally's birth sign is Aries not Virgo. She wears a plaid skirt not a striped one. She
    prefers an olive to a nut. Are her eyes hazel or brown?

See answer No. 82

Target Practice

You have a maximum of 6 bullets but you are required to score exactly 100. Can you do it on this unusual target?

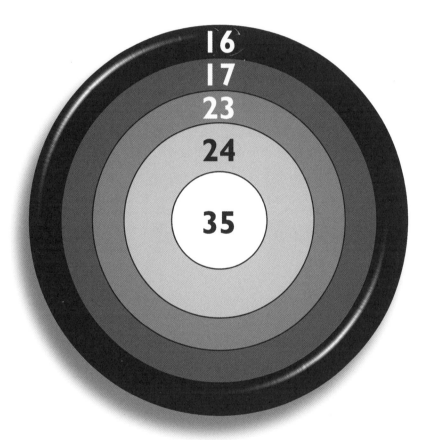

See answer No. 11

Incorrect Use of the Calculator

A schoolboy was cheating with his homework. The teacher had asked the class to calculate the answers without the use of a calculator. He came to one sum that was in the form of (a + b) divided by c = Answer. When he used his calculator he got some very strange results because he had not completed the sum within the brackets first. His result should have been 16, but his first attempt gave him an answer of 24. He thought this might not be correct so he reversed the order of entering "a" and "b", which gave him a more unlikely result of 40. What were the values of the 3 letters?

See answer No. 66

Ballet or Opera?

A survey discovers that 8 more people like opera only than ballet only. One more person likes the theatre only than the opera only; 44 people taking part in the survey do not like opera at all; 67 people do not like ballet; 9 more people like theatre only than ballet only; 14 people like both ballet and opera but not theatre, and 26 like both opera and theatre but not ballet; 78 people like the opera.

   a)   How many people like opera only?
   b)   How many people like ballet only?
   c)   How many people like theatre only?
   d)   How many people like all three?
   e)   How many people took park in the survey?

See answer No. 59

Magic Squares with Dominoes

A set of dominoes consists of 28 tiles with spots ranging from 0–0 to 6–6. Can you complete the 5 x 5 square by discarding the 0–5, 0–6 and 1–6 tiles so that it forms a magic square that has 30 spots in each horizontal row, vertical column, and major diagonals?

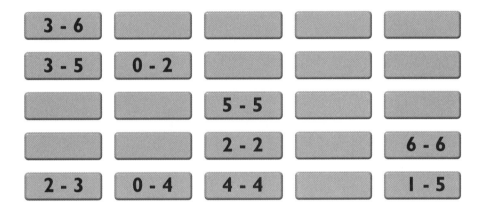

See answer No. 126

Ham It Up

There is a choice of cheese, ham, or salad, or any combination of the 3, for sandwich fillings. The same number of people have ham only as have all 3 of the fillings. Twice as many have both ham and salad but no cheese as have all 3 fillings; 14 people do not have cheese. Three more people have cheese only as salad only; 8 people have both ham and salad but no cheese; 12 people do not have salad; and 14 people do not have ham.

a) How many people have all 3 fillings?

b) How many people have cheese only?

c) How many people have both cheese and salad but no ham?

d) How many people have only one filling?

e) How many people have only 2 fillings?

See answer No. 22

Card Sharp

In a card shop five times as many people buy a card only as wrapping paper only. One more person buys both a card and a bow but no wrapping paper as a bow only. Twice as many people buy both a bow and wrapping paper but no card as buy all three. Twelve people do not buy a card and 18 do not buy wrapping paper; 15 people buy a card only and 4 people buy all three, and 30 people do not buy a bow.

a) How many people buy 1 of the items only?

b) How many people buy 2 of the items only?

c) How many people buy wrapping paper only?

d) How many customers are there in total?

e) How many people buy a bow, wrapping paper, but no card?

See answer No. 57

Party Poser

At a party twice as many people wear a watch only as both a ring and a bracelet but no watch. there are 45 people who wear both a watch and a ring but no bracelet; 79 people do not wear a bracelet and 53 do not wear a ring. Eight more people wear a ring only as a bracelet only, 10 people wear both a ring and a bracelet but no watch. Three times as many people wear a watch and a ring but no bracelet as wear all three.

a) How many people wear a watch and a bracelet but no ring?
b) How many people wear all 3?
c) How many people wear a bracelet only?
d) How many people wear a watch only?
e) How many people were wearing only 2 of the 3?

See answer No. 90

Reader Riddle

In a survey twice as many people said they read both the sport and the TV pages but not the headlines, as read the TV pages only. Four more people read the headlines only as the sport only: 12 people read the TV pages only: 2 more people read the headlines only as read both the headlines and the TV pages but not the sport. One more person reads both the headlines and the sport as reads all 3. 66 people do not read the headlines, and 112 do not read the TV pages: 34 people read the headlines only.

a) How many people read only 1 section?

b) How many people read the sport and the TV but not the headlines?

c) How many people read only 2 sections?

d) How many people read all 3 sections?

e) How many people took part in the survey?

See answer No. 120

DIY Dilemma

In a home-improvement store selling only paint, wallpaper, and tiles, twice as many people buy wallpaper only as tiles only. Three more people buy paint only as wallpaper only. One more customer buys both paint and wallpaper as buys tiles only. Twice as many people buy tiles only as both tiles and wallpaper. Eighteen people do not buy wallpaper and 14 do not buy paint. The same number of customers buy both tiles and wallpaper only as buy all three; 5 people buy both wallpaper and paint only.

a) How many people buy wallpaper only?

b) How many people buy all 3?

c) How many customers are there in total?

d) How many people buy only 2 of the 3?

e) How many people buy paint only?

See answer No. 41

Conveyor Belt Conundrum

A customer at the supermarket places 7 items on the conveyor belt. The last item placed on it is a cake. The milk is placed immediately before the biscuits. The fruit juice is placed immediately after the apples. The soup is placed immediately before the milk. There are 2 items between the bread and the biscuits, and 2 between the soup and the apples. The bread is placed immediately before the cake.

a) Which item is placed on the conveyor belt immediately before the bread?
b) Which item is placed on the belt first?
c) Which is the item placed fourth?
d) In which place is the milk placed?
e) Which item is placed after the milk but before the apples?

See answer No. 104

Seven Sequence

Seven boys are having a race. George finishes 2 places behind Liam. Jack finishes behind Alex. Clive finishes after David but before Liam. Ben finishes 2 places ahead of Alex, Liam is the third to finish.

a) **In which position does Ben finish?**

b) **Who wins the race?**

c) **Who finishes fifth in the race?**

d) **Who finishes immediately after George?**

e) **Who finishes last?**

See answer No. 17

Lost Elevator

The office elevator visits 6 floors. Sophie, Ted, James, Claudia, Joanne, and Mark each want to go to a different floor. The elevator is going up. Joanne gets out after Sophie but before James. Mark gets out 1 stop after Ted and 3 stops before James. James is not the last person to get out of the elevator.

a) **Who gets out of the elevator first?**
b) **Who gets out of the elevator last?**
c) **Who gets out just after Mark?**
d) **Who is the fourth person to leave the elevator?**
e) **Who is the second person to leave the elevator?**

See answer No. 114

Who's Where?

In a 6-storey block of apartments Ms. Smith lives 3 floors above Mr. Thomas. Ms. Smith lives 2 floors above Ms. Baker. Mr. Brown lives just above Ms. Smith. Mr. Thomas lives above Mr. Ridge but below Ms. Baker.

a) On which floor does Mr. Lloyd live?

b) Who lives on the top floor?

c) Who lives directly above Mr. Thomas?

d) Who lives on the bottom floor?

e) On which floor does Ms. Smith live?

See answer No. 61

Waiting Game

Six people are waiting in the bus queue. Mr. Cole is standing ahead of Ms. Fielding. Mr. Barnes is three places behind Mr. Sharp. Mr. Jones stands behind Ms. Richards. Mr. Barnes stands immediately behind Ms. Fielding.

a) **Which position in the queue does Mr. Cole hold?**

b) **Who is at the front of the queue?**

c) **Who is last in the queue?**

d) **Who is third in the queue?**

e) **Which position in the queue does Ms. Richards hold?**

See answer No. 100

Unlucky Sailor

A sailor returned to his ship after a few days of shore leave, telling his crew mates how lucky he was to get something. Without hesitation his crew mates threw him overboard for mentioning it as being lucky. What was it that he had said?

Clues

1. It was a bit of good fortune, but it did not involve money or wealth.
2. The sailor was really very pleased before he got back to the ship, and if he had thought about it, he may have phrased his tale differently.
3. His other friends who witnessed what had occurred on land were pleased for him.
4. It was a sporting achievement.

See answer No. 110

Square Grids

a) How many squares can you see in this drawing?

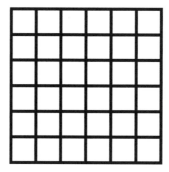

b) Can you see any easy way to count the squares for any size matrix of squares?

See answer No. 36

Do You Need a Computer?

What comes next in this series?

?

See answer No. 47

Equal Shapes

Can you divide this matrix along the lines into four parts that contain both a triangle and a star? Each part must be exactly the same shape and size, but the triangle and star positions will vary.

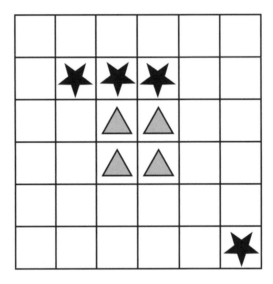

See answer No. 112

Double Square

If you make two straight cuts on a piece of paper shaped as a cross (see below), can you rearrange the cut cross to form 2 squares?

See answer No. 16

Triangles

Which triangle has the larger area?

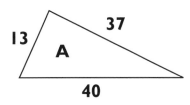

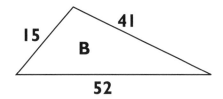

See answer No. 58

Handyman

Alf had been asked by his wife to put tiles on an old square table in the garden. This would smarten the table up and they could then use it when their friends came over for a barbecue. Alf made a drawing and showed his wife. She said that it would look better if the square tiles were 3/4 of an inch smaller. Alf said that that would require 250 more tiles. What was the original tile size and how many tiles would he now have to fit?

See answer No. 107

Divided Square

This field measures 177m x 176m. It has been split up into 11 squares that exactly equal the total area. The squares are only roughly drawn to scale. All new squares are in whole yards. Can you calculate the size of each square?

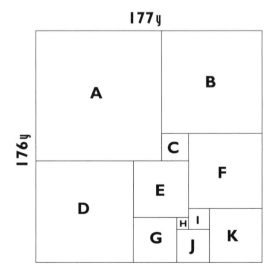

See answer No. 44

Japanese Door Sign

This sign was printed on a door in Japan. What does it mean?

See answer No. 111

The Prince of Wales' Feathers

A lady was walking down a street in London when three feathers fell from above and knocked her unconscious. How could this be explained?

See answer No. 18

Complex Division

Using each of 9 digits only once, find a number where the first digit is divisible by 1. The first two digits must be divisible by 2. The first 3 digits must be divisible by 3 etc. up to 9 digits that must be divisible by 9.

See answer No. 68

Word Stepping

Start with a word on the top line, move down, sideways, diagonally, or up to the second word, then tack the next word on the end to form a new or compound word, and so on. Use 16 words to form 15 compound words.

Eg. Whim - Per
Per - Son

Whim	Break	Sup	Penny	Currant
Fast	Per	Piece	Bun	Weight
Net	Son	Time	Meal	Bath
Her	Step	Table	Cock	Pit
Father	Ring	Value	Ball	Boy
Let	Worm	Screw	Cap	Size

See answer No. 46

But We Need the Beds!

The need for hospital beds is critical, but for some strange reason patients who have recovered enough to leave hospital refuse to go for 2 more days. The doctors permit this. Why?

Clues

1. This is based on fact.
2. The patients would not be at risk if they left when the doctor originally told them.
3. If the doctor said they could leave the day before they would have gladly done so.
4. It was a Saturday in Dublin (Eire).

See answer No. 85

The Panel Game

Three people were hidden behind a screen and they were asked to make statements about themselves and the others behind the screen. They had to include only one lie and then the contestants had to identify the person who had a yellow marker. We will call the three people A, B, & C. Which one of them had the marker?

'A' and 'C' said   'I have the marker.'
'A', 'B' & 'C' said   'B does not have it.'
'B' & 'C' said   'C does not have it.'
'A' said   'Neither A nor C has it.'
'B' said   'A does not have it.'

See answer No. 119

Spider's Logic

Wⁿhat is the value of the end web?

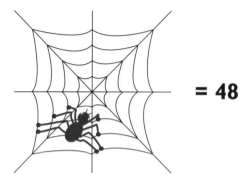

= 48

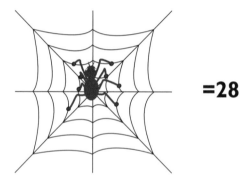

=28

= ?

See answer No. 67

The Barber of Seville

The Barber of Seville shaves all the men living in Seville. No man who lives in Seville is permitted to shave himself. The Barber of Seville lives in Seville. The Barber of Seville does not leave the city and no one from outside enters the city. Who then shaves the Barber of Seville?

See answer No. 29

Missing Vowels

Replace the missing vowels in the following groups to form 4 associated words. What are they?

1. KRT   CRQT   LCRSS   NGLNG
2. LDL   SPTL   CRT   CTLRY
3. GRNGG   PRCT   KMQT   GRPFRT
4. NDL   PNT   PNCK   MRNG
5. GRLL   LPRD   RNDR   PRCPN

See answer No. 54

Mixed Letters

Rearrange the order of letters in each group to form a word. What are the four connected words?

1.	HORTMABO	GELONU	THICENK	DOMOBER
2.	ROSSSCIS	FINKNEEP	WESTZEER	WORKCECRS
3.	RYLRO	OCHCA	TACHY	PORTCHEELI
4.	BEGBACA	YERCEL	TREEBOTO	KNIPUMP
5.	LISK	NINEL	NOLNY	NOTCTO

See answer No. 42

What am I?

1. I am a sauce. Change one letter and I am a tomb. Change another letter and I am a fireplace. Change another letter and I am a fruit. Change one final letter and I am a chart. What was I and what did I become?

2. I am also called a shop. Change one letter and I am a bird. Change another letter and I am bare. Change another letter and I am the beginning. Change one final letter and I am neat. What was I and what did I become?

See answer No. 121

Grid Codes

1.

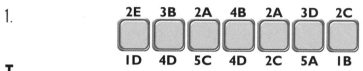

The word-frame above, when filled with the correct letters, will create the name of a planet. The letters are arranged in the coded square below. There are 2 alternatives to fill each square of the word-frame, one correct the other incorrect. What is the planet?

	A	B	C	D	E
1	B	Y	X	S	C
2	T	W	R	X	J
3	H	A	G	M	Q
4	V	I	G	U	F
5	E	O	P	D	Z

2.

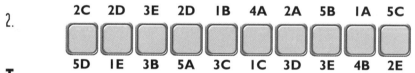

The word-frame above, when filled with the correct letters, will create the name of a state of the U.S. The letters are arranged in the coded square below. There are 2 alternatives to fill each square of the word-frame, 1 correct the other incorrect. What is the state of the U.S.?

	A	B	C	D	E
1	E	F	O	G	A
2	R	V	M	I	S
3	Q	L	Y	K	N
4	B	I	T	H	D
5	T	P	A	C	U

Grid Codes/continued

3.

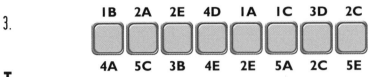

The word-frame above, when filled with the correct letters, will create the name of a mountain range. The letters are arranged in the coded square below. There are 2 alternatives to fill each square of the word frame (one correct, the other incorrect). What is the mountain range?

	A	B	C	D	E
1	I	P	R	F	S
2	Y	W	E	Q	N
3	M	R	J	C	T
4	D	F	L	E	D
5	E	H	A	G	S

4.

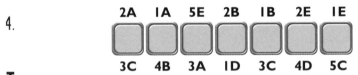

The word-frame above, when filled with the correct letters, will create the name of a country. The letters are arranged in the coded square below. There are 2 alternatives to fill each square of the word-frame (one correct, the other incorrect). What is the country?

	A	B	C	D	E
1	A	I	D	G	K
2	C	R	V	W	U
3	T	Y	B	J	F
4	N	E	S	O	T
5	H	Z	M	P	L

Grid Codes/continued

5.

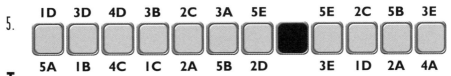

ID 3D 4D 3B 2C 3A 5E ⬛ 5E 2C 5B 3E

5A IB 4C IC 2A 5B 2D ⬛ 3E ID 2A 4A

The word-frame above, when filled with the correct letters, will create the name of a film star. The letters are arranged in the coded square below. There are 2 alternatives to fill each square of the word-frame (one correct, the other incorrect). Who is the film star?

	A	B	C	D	E
I	F	A	S	R	H
2	A	K	E	L	V
3	L	H	X	I	G
4	E	N	C	M	T
5	T	R	B	O	D

See answer No. 86

Empty Squares

1. Should A, B, C, or D replace the empty squares in the grid?

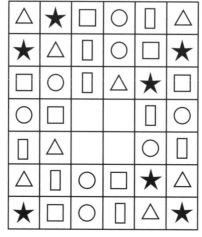

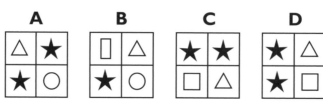

Empty Squares/continued

2. Should A, B, C, or D replace the empty squares in the grid?

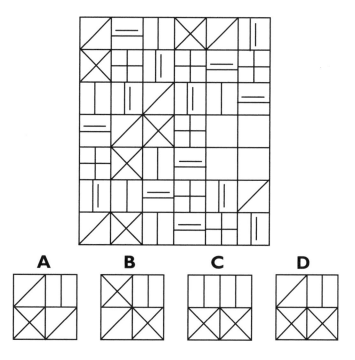

A   B   C   D

Empty Squares/continued

3. What number should replace the question marks?

5	2	1	9	7					2	8	9	0	5					6	1	3	2	4			
1	6	8	8	4	G	A	I		3	2	0	2	6	D	F	B		2	3	7	1	7	C	H	E
6	9	8	0	0					6	1	3	9	3					?	?	?	?	?			

4. What number should replace the question marks?

4	2	5	3	1					7	0	1	3	6					1	9	9	0	8			
3	8	2	8	7	F	D	A		2	2	1	4	5	H	I	E		5	7	4	6	7	G	B	C
8	0	1	7	7					9	1	3	8	6					?	?	?	?	?			

See answer No. 4

Longest Word

1. Move from square to adjacent square to discover the longest possible flower name in the grid. What is it?

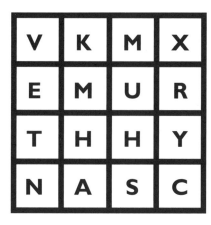

2. Move from square to touching square to discover the name of a famous musical. Use as many of the letters as possible. What is it?

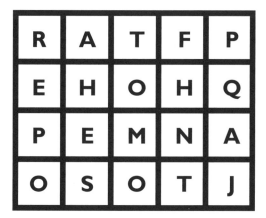

See answer No. 34

Number Links

1. Rearrange the tiles and place them touching one another in a 3 x 3 square, so that all adjacent letters are the same. If placed correctly, a type of dinosaur will be read around the outer rim. What is it?

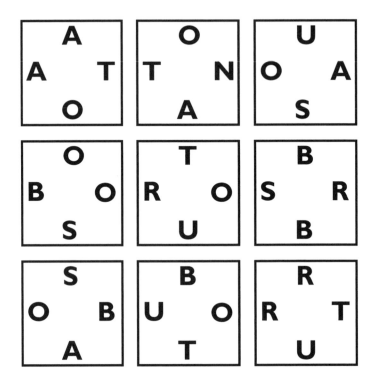

Number Links/continued

2. Rearrange the tiles and place them touching one another in a 3 x 3 square, so that all adjacent letters are alphabetically consecutive. If placed correctly, the name of a city in the U.S.A. will be read. What is it?

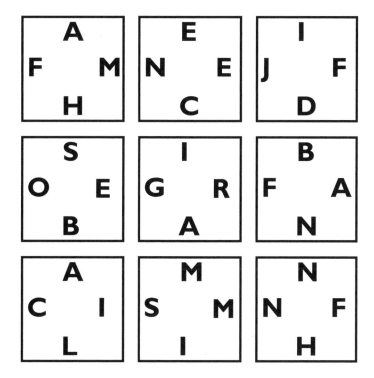

See answer No. 118

Unusual Safes

1. Here is an unusual safe. Each button must be pressed once only in the correct order to reach OPEN. The number of moves and the direction is marked on each button (1S would mean move one space south, etc.). Which button should you press first?

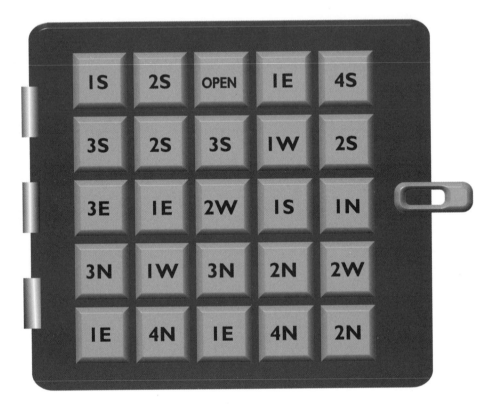

Unusual Safes/continued

2. Here is an unusual safe. Every single button must be pressed once only in the correct order to reach OPEN. The number of moves and the direction is marked on each button (3C would mean 3 moves clockwise, and 1A would mean one move anti-clockwise). Which button should you press first?

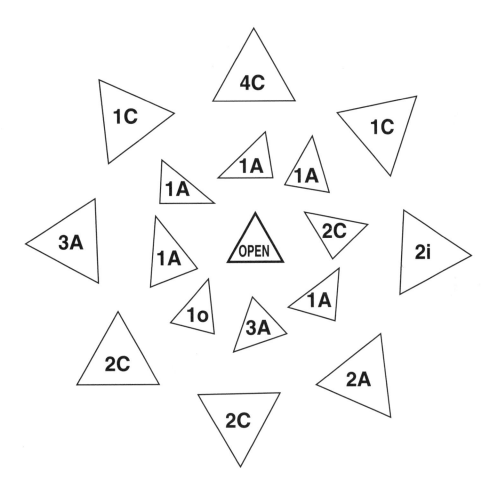

Unusual Safes: Codecracker

1. Place a letter in the middle of the diagram so that a word can be read along each straight line. What are the 4 words?

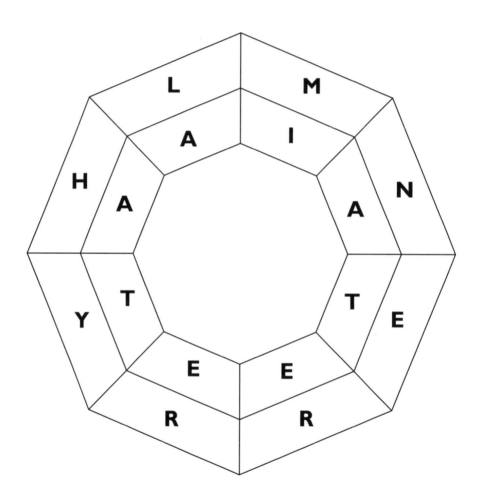

Unususal Safes: Codecracker/continued

1. Place a letter in the middle of the diagram so that a word can be read along each straight line. What are the 4 words?

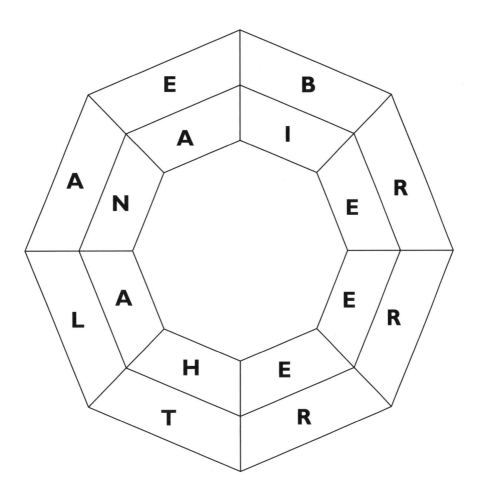

See answer No. 79

Blank to Blank

Can you fill in the blanks in these sentences? The second blank is an anagram of the first.

1. Lucy made a note in her ———— of the day she visited the ————.

2. At the fruit shop the lady picked up a ———— which was very ————.

3. The father told his son a ————, but his mother said it was too ———— at night.

4. The chef turned the heat on the cooker down to ———— the ———— boiling over.

5. The woman asked her husband what he wanted to ———— with his cup of ————.

See answer No. 7

Code Wheel

Place each of the letters in the middle of this circle in a box around the edge. If correctly placed, the letters will be the first letter of 1 word and the last of another, and 8 6-letter words will be read. What are they?

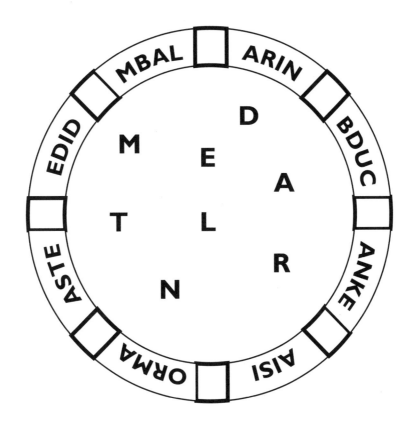

See answer No. 24

Pentagons

1. Write each of the following 5-letter words clockwise around a pentagon. Where 2 pentagons join, the 2 facing segments must contain the same letter. Some letters are given for you. How should the completed diagram look?

OCCUR, KNACK, AGREE, ODOUR, IGLOO, CHINA, SKATE, BUILD, STAMP, TASTE.

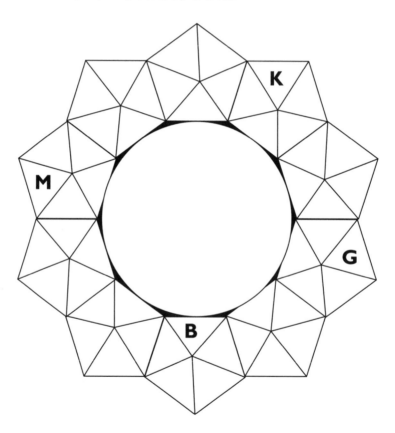

See answer No. 60

Pentagons/continued

2. Write each of the following 5-letter words clockwise around a pentagon. Where 2 pentagons join, the 2 facing segments must contain the same letter. Some letters are given for you. How should the completed diagram look?

ALERT, EBONY, USHER, ALTAR, ANGLE, ACUTE, START, BEACH, EAGER, GRILL.

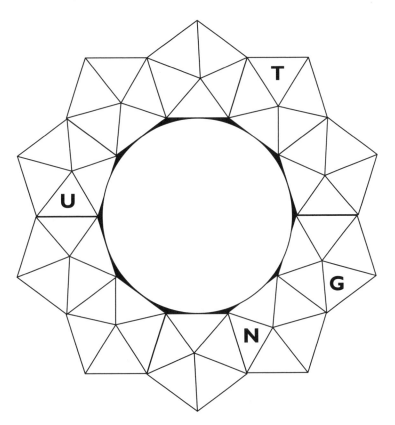

See answer No. 60

Columns

1. Rearrange the order of the given words and place 1 word on each row of the grid. If the words are in the correct order, the name of a fruit can be read down each of the shaded columns. What are the 3 fruits?

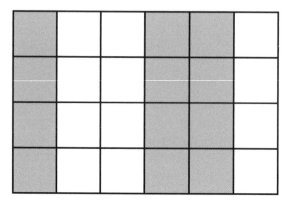

AUTUMN PEOPLE RHYMES EYELID

Columns/continued

2. Rearrange the order of the given words and place 1 word on each row of the grid. If the words are in the correct order, the name of a country can be read down each of the shaded columns. What are the 2 countries?

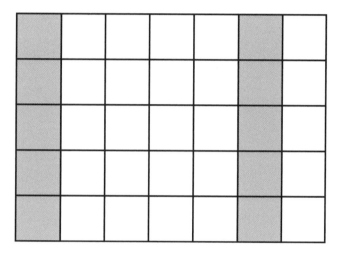

LENIENT ANGELIC YASHMAK INFANCY THOUGHT

Columns/continued

3. Rearrange the order of the given words and place 1 word on each row of the grid. If the words are in the correct order, the name of a Formula 1 racing team can be read down each of the shaded columns. What are the 2 teams?

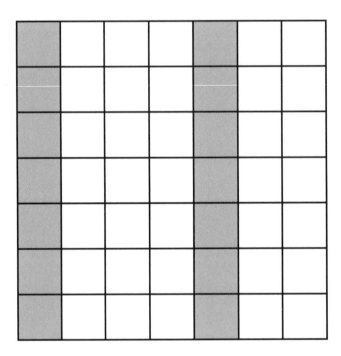

LIBERTY RADIANT EMPEROR
MINDFUL CLOSEST ADMIRAL NOURISH

Columns/continued

4. Rearrange the order of the given words and place 1 word on each row of the grid. If the words are in the correct orde,r the name of a sweet food can be read down each of the shaded columns. What are the 2 foods?

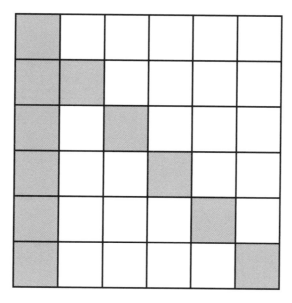

EMPIRE ROBUST FULFIL LADDER TALENT INFORM

See answer No. 124

Letter Grid

1. Use the letters given to complete the square so that 4 other words can be read down and across. What are the words?

S	P	L	I	T
P				
L				
I				
T				

AEEELLLSSSSSTTUU

Letter Grid/continued

2. Use the letters given to complete the square so that 4 other words can be read down and across. What are the words?

F	A	T	A	L
A				
T				
A				
L				

AAAADEEMMOORRYYZ

See answer No. 49

Pyramid Letters

1. Use the letters given to complete the pyramid so that words can be read down and across. Across there will be one 7-letter word, one 5-letter word, and one 3-letter word. Running down will be one 4-letter word and two 3-letter words. What are they?

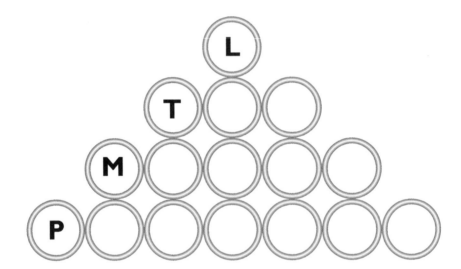

A E G I N N N O O R U Y

Pyramid Letters/continued

2. Use the letters given to complete the pyramid. Words can be read down and across. Running down will be 2 words of 3 letters and 1 word of 4 letters. Across there is 1 word of 3 letters and 1 of 5 letters. What are the words?

S N O W M A N

A A E H K L O T T

See answer No. 75

Merged Words

1. The names of 3 birds are merged together here. What are they?

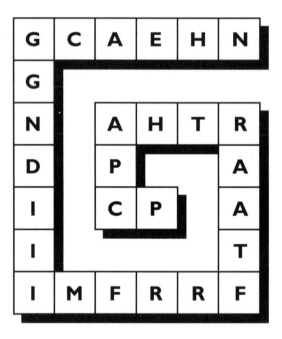

Merged Words/continued

2.   The names of 3 animals are merged together here. What are they?

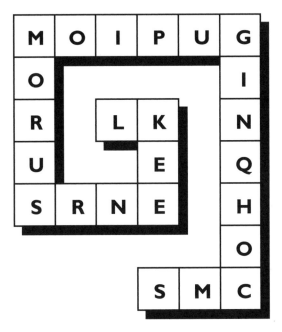

See answer No. 105

The Diners

Mr and Mrs A invited 3 couples around for dinner. They were Mr and Mrs B, Mr and Mrs C, and Mr and Mrs D. The seating arrangements were such that one couple sat apart. Can you figure out which couple this was given the following:

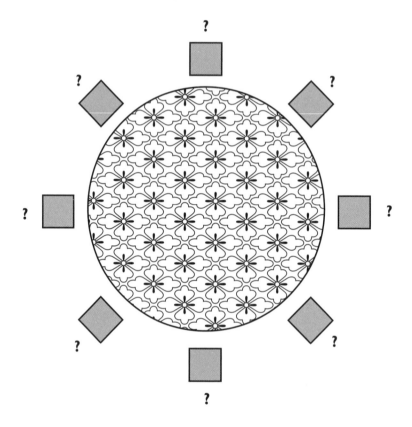

- The person opposite Mrs. A was the man who sat on Mr B's left.
- The person to the left of Mrs. C was a woman who sat opposite Mr D.
- The person to the right of Mr. D was the woman who sat opposite the woman two seats to the left of Mr. A.

See answer No. 73

Scared

Late one evening a woman returned alone to her home. She was very tired so got straight into bed without switching on any lights. Within minutes she had a horrible feeling she was not alone. She opened her eyes to see a dark shadow over by the window. The woman was too frightened to scream or even to move. Almost immediately the shadow started to move toward her. She panicked and grabbed for the light switch. The light shining towards him seemed to frighten him as much as the woman, and he ran out of the door without physically harming her. The woman shut the door behind him and got back into bed. Although the woman was petrified she did not report the incident and she did not worry that he would be back to attack her. Why was this?

See answer No. 65

Old Grandpa!

Grandpa was aged between 50 and 70 years. He told his friends,

> 'Each of my sons has as many sons as brothers, and the combined number of sons and grandsons is precisely the same as my age.'

How old was Grandpa?

See answer No. 76

Unscramble the Bull

A bull in a field eats an unexploded bomb. Can you unscramble these 4 words and determine what word best fits the situation before the bull explodes?

B E E I L R R T
A A C C H I O P R S T T
A A B B E I L M N O
E E I L O P S V X

See answer No. 20

Who Am I?

I was born in 1859 in Edinburgh, Scotland.
I wrote about a detective and an arch-criminal who was a professor.
I was a medical doctor, just like one of the main characters in my books.
I was made a knight for my work in field hospitals in the South African (Boer) War.

See answer No. 109

What Title Do I Have?

I am a sapphire-blue gem stone found in India.
Originally I was bought by Louis XIV in 1668 as a part of the French crown jewels but was later cut up and called the French Blue.
I was then stolen and cut up again.
I am named after the person who bought me in 1830.
For many years you could see me at the Smithsonian Institute (Washington DC).

See answer No. 51

Who Was I?

I was originally called ERIK WEISZ and was born in Hungary.
My father was a rabbi who emigrated from Hungary to the USA.
I was a conjuror and trapeze artist.
I was more famous for my feats of strength and life-threatening entrapments.

See answer No. 64

Word Squares

Insert the given letters in the blank spaces so that the same words are read horizontally and vertically.

A)

V	A	L	E	T
A				
L				
E				
T				

BBEEEENOOSSTTVVY

Word Squares/continued

B)

C	H	E	S	T
H				
E				
S				
T				

DEEELLLLOOOORRSSV

See answer No. 56

Missing Letters

The missing letters in each of the rows below will form words when identified and rearranged. Each group has a connection, which is given as an aid. Can you identify the answers ?

1. Scientists
   a) BDEFGHIJMOQRSTUVWXYZ
   b) ABFGHJKMNOPQRSTVWXYZ
   c) ABCFGHJKLMPQRTUVWXYZ

2. Planets
   a) ABCDFGHKLMNOQSVWXYZ
   b) ABCDEFGHIJKMNQRSVWXYZ
   c) BCDEFGHIJKLMOPQVWXYZ

3. Composers
   a) ABCDEFGHJKMNOPQRUVWXY
   b) BCEFGHIJLMNPQSTUWXYZ
   c) BCDFGIJKNOPQSTUVWXYZ
   d) BCDEFGHIJKLNPQSUVWXY

See answer No. 83

Link Words

What word links the words on the left with the words on the right?

a] clothes   box
    hobby   hair
    war   play
    work   race

b] wine   ware
    shot   eye
    hour   paper
    stained   house

c] dog   time
    television   boat
    fashion   ground
    car   off

d] day   table
    life   frame
    mean   piece
    over   warp

See answer No. 97

Currencies

By using one letter from 3, 4, 5, or all 6 adjoining rows can you find the following ?

a) 2 items of currency of 3 letters.
b) 4 items of currency of 4 letters.
c) 4 items of currency of 5 letters.
d) 3 items of currency of 6 letters.

F K D P Y S C M B
R J H E O H E A C
A O L S U E N R H
O D N L E C K T L
C A D D Q E U G
M R F A G L U K W

Remember you may not skip a line.

See answer No. 127

Word Oddities

1. Can you think of 2 words that have each of the vowels appearing once only in them, but in alphabetical order?
2. Can you name an 8-letter word with all of the vowels in it?
3. Can you name 2 words of 9 letters with 2 pairs of double Os in them?
4. Can you name a word with 6 Is in it?
5. Can you think of 2 words that contain 4 Us?

See answer No. 6

Circular Crossword

Answer the clues given in the main circles and enter the letters of your answer in the 6 outer circles. Start at the pointer and move in the direction of the indicator. When completed, the shaded circles will give you a word or phrase when they are read from the top of the puzzle toward the bottom of the puzzle.

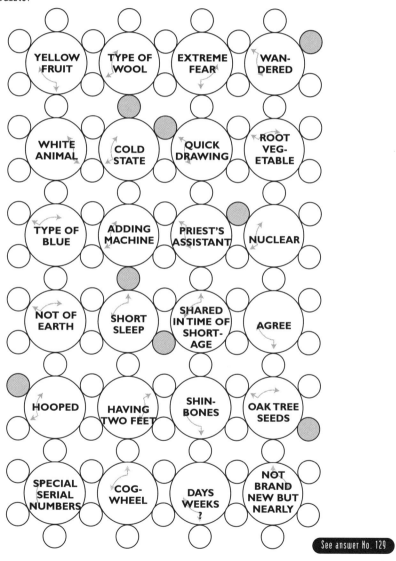

YELLOW FRUIT
TYPE OF WOOL
EXTREME FEAR
WAN-DERED

WHITE ANIMAL
COLD STATE
QUICK DRAWING
ROOT VEG-ETABLE

TYPE OF BLUE
ADDING MACHINE
PRIEST'S ASSISTANT
NUCLEAR

NOT OF EARTH
SHORT SLEEP
SHARED IN TIME OF SHORT-AGE
AGREE

HOOPED
HAVING TWO FEET
SHIN-BONES
OAK TREE SEEDS

SPECIAL SERIAL NUMBERS
COG-WHEEL
DAYS WEEKS ?
NOT BRAND NEW BUT NEARLY

See answer No. 129

What the ??

What letter should replace the question mark in these series?

1. **J A S O ? D**

2. **A T G ? L V L S S C A P**

3. **M V E M J ? U N P**

4. **U O U E H R ?**

5. **Z X C V B ? M**

6. **N W H O I ? E I I**

What the ??/continued

7. If 6 people were born under the star sign of Libra, 3 under the sign of Aries and 4 under the sign of Leo, how many were born under the star sign of Gemini?

8. If Paula gives Jenny $6 they will both have the same amount of money. If Jenny gives Paula $6, Paula will have 7 times as much as Jenny. Who has what?

9. Where does it appear that 1089 + 2568 equals 15753, and 6621 + 8512 equals 3457?

10. If $L + K = 5$, $P + T = 3$ and $N + Z = 6$, what does $A + D = ?$

See answer No. 89

Code Wheels

In this system of intermeshing wheels, the large wheel rotates clockwise and drives the other wheels. The number of teeth for each wheel is given, and each tooth is represented by a letter. Can you calculate what words, or letters, will be at the indicator positions if:

a) The large wheel rotates through 3 revolutions?
b) The large wheel rotates through 4.5 revolutions?
c) The large wheel rotates through 5 revolutions?
d) The large wheel rotates through 6 and $1/3$ revolutions?

18T   16T   12T   8T

See answer No. 115

More Code Wheels

In this system of intermeshing wheels, the large wheel rotates clockwise and drives the other wheels. The number of teeth for each wheel is given, and each tooth is represented by a letter. Can you calculate what words, or letters, will be at the indicator positions if:

a)   The large wheel rotates through 2 revolutions?
b)   The large wheel rotates through 5 revolutions?
c)   The large wheel rotates through 7 revolutions?
d)   The large wheel rotates through 4.5 revolutions?

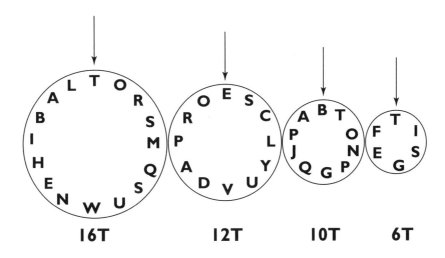

| 16T | 12T | 10T | 6T |

See answer No. 91

Logical Deduction Answers

1 **The Rector Total**

S = 0	T = 6
A = 1	O = 7
C = 2	R = 8
P = 3	L = 9
E = 4	

2 **Washers**

πr^2 = Area of a circle

The area of a tarus (washer-shaped item) = πh^2

3 **Lateral Thinking with Matchsticks**

Ans. 1   Ans. 2

```
| | | |  |  | | | |  =9        | | |  ✕  | | |  =9

  |       |                       ✕       ✕

| | | |  |  | | | |  =9        | | |  ✕  | | |  =9
```

4 **Empty Squares**

1. D. The five symbols run along the top row, then along the second row, etc.
2. B. The six symbols run down the first column, then spiral round to the middle of the grid.
3. 85426. Replace the letters with their alphabetical value (A=1, B=2, etc.), then add them to lines 1 and 2 of the previous box to create line 3.
4. 76652. Replace the letters with their alphabetical value, then deduct them from the sum of lines 1 and 2 of the previous box to create line 3.

5 **Numerical Links**

425. Alphabetical value of each letter is squared and added together by line.

6 **Word Oddities**

1. Facetious, Abstemious
2. Equation
3. Footstool, Foolproof
4. Indivisibility
5. Tumultuous, Unscrupulous

7 **Blank to Blank**

1. DIARY and DAIRY.
2. PEACH and CHEAP.
3. TALE and LATE.
4. STOP and POTS.
5. EAT and TEA.

8 **So You Think You're Good at Math?**

Just turn it upside down to get 81 + 19 = 100

9 **Word Connections**

1. Cars: Lotus, Fiat, Ford, Saab, Audi.
2. Tennis players: Hingis, Cash, Henman, Bates.
3. Musical instruments: Tuba, Harp, Lyre, Drum.
4. Breeds of Dog: Collie, Basset, Boxer, and Beagle.
5. Gems: Amber, Opal, Coral, Pearl.

10 Friend
The friend was a man.

11 Target Practice
100 = 16 + 16 + 17 + 17 + 17 + 17, the only solution

12 Train the Train Driver

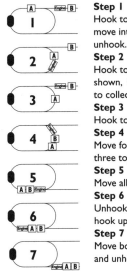

Step 1
Hook to load B, reverse to A, and move into position shown and unhook.
Step 2
Hook to A and unhook in position shown, then go through the tunnel to collect B.
Step 3
Hook to B and reverse.
Step 4
Move forward to connect all three together.
Step 5
Move all three to position shown.
Step 6
Unhook train, go around loop, and hook up to load A.
Step 7
Move both loads to position shown and unhook B.
Step 8
Reverse load A into position shown.
Step 9
Unhook train and go around the loop to position shown.
Step 10
Collect load B and reverse toward load A.
Step 11
Move load B to position shown, and return train to the original position.

13 Miss Punctual
She had a mains-operated electric clock. The power to her house had been cut off for just over 30 minutes while she slept. As a result the buzzer was late .

14 A Lewis Carroll Gem
I

15 The Books
It was a library and she had to pay a fine for being overdue.

16 Double Square

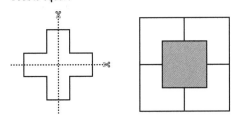

17 Seven Sequence
a) Fourth
b) David
c) George
d) Alex
e) Jack

18 The Prince of Wales' Feathers
It was a pub sign that had fallen. The place was called The Prince of Wales' Feathers.

19 Look for the Simple Solution
Zero, (t - t) = 0 and anything multiplied by zero is zero.

20 Unscramble the Bull
Abominable (A bomb in a bull)
Other words were: terrible, catastrophic, explosive.

21 Magic Squares

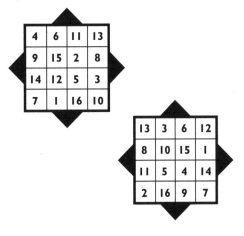

22 Ham it Up
a) 4 b) 5 c) 7 d) 11 e) 18

23 Treasure Island
GR.

1.	Statue of Liberty	SOL
2.	Eiffel Tower	ET
3.	Queen	Q
4.	Kermit	K
5.	Fort Knox	FK
6.	Joseph	J
7.	Isaac Newton	IN
8.	Humphrey Bogart	HB
9.	A.D.	AD
10.	Prime Minister	PM
11.	C in Roman (Century)	C

24 Code Wheel
Marina, Abduct, Tanker, Raisin, Normal, Lasted, Decide, Embalm.

25 More Lateral Deduction

(1–5) In each case the Roman numerals are added up.

1. 101
2. 501
3. 10
4. 150
5. 1001
6. 24. Each vowel is worth 8.
7. 6. Each vowel is worth 3.
8. 24. Each vowel is worth 12.
9. 63. Each vowel is worth 21.
10. 144. Each vowel is worth 36.
11. 35. Each consonant is worth 7.
12. 55. Each consonant is worth 11.
13. 135. Each consonant is worth 27.
14. 42. Each consonant is worth 14.
15. 140. Each consonant is worth 35.
16. 5. Each vowel is worth 1, each consonant is worth 2.
17. 14. Each vowel is worth 5, each consonant is worth 1.
18. 22. Each vowel is worth 4, each consonant is worth 6.
19. 22. Each vowel is worth 3, each consonant is worth 4.
20. 27. Each vowel is worth 2, each consonant is worth 7.
21. 134. Alpha positions of first word plus alpha positions of second word.
22. 15. Alpha positions of first word minus alpha positions of second word.
23. 4240. Alpha positions of first word multiplied by alpha positions of second word.
24. 196. Alpha positions of first word plus alpha positions of second word multiplied by 2.
25. 45. Alpha positions of first word plus alpha positions of second word divided by 2.
26. 82. Alpha positions of first word minus alpha positions of second word multiplied by 2.
27. $92\frac{1}{2}$. Alpha positions of first word plus half alpha positions of second word.
28. 243. Alpha positions of first word plus twice the alpha positions of the second word.
29. 2956. Alpha positions of first word squared plus alpha positions of second word.
30. 1957. Alpha positions of first word squared minus alpha positions of second word squared.

26 The Rabbit Family
8 male plus 6 female.

27 Random Chance
None. If you get three right, the fourth will also be right.

28 Holes In Paper
32 : It follows the formula 2^{n-1} where n equals the number of folds.
$$2^{6-1} = 2^5 = 32$$

29 The Barber of Seville
The Barber of Seville is a woman.

30 Not Really a Premonition
He was sky-diving and his parachute failed to open.

31 Equal Segments
1)
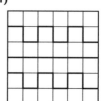
2)
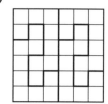

32 Fact or Fiction
February was added to the calendar a little later and nobody knows when he actually did it. The early calendar only had 10 months.

ANSWERS

33   That & This

Let 'This' = x   then 'That' = 8x

So x + 8x = $3x^2$ = 9x

So x = 3 = 'This'

& 'That' = 24

34   Longest Word

1.   Chrysanthemum.

2.   Phantom of the Opera.

35   A Waiter's Lot

a) 14 b) 4 c) 18 d) 7 e) 8

36   Square Grids

a)   91

b)   The easy way is to start with a square of 1 x 1, then 2 x 2 and so on until you see the sequence.

1 sq = 1

2 sq = 5 (1 + 2 x 2)

3 sq= 14 (5 + 3 x 3)

4 sq = 30 (14 + 4 x 4)

5 sq = 55 (30 + 5 x 5)

6 sq = 91 (55+ 6 x 6)

7 sq = (55+ 7 x 7) = 104

37   Strange But True!

They suggested drawing a line on the tall man's chest level to where the short man stood. Any shot above that line was not to count.

38   Complex Numbers & Letter Grids

1.   48. (A x B) − (C x D) = ef

2.   M x 2A − B + C = D (Use alphabetical values of each letter.)

3.   21. a + b + e + f = cd

4.   32. (a x b) divided by c = d

39   The Detective Booking Clerk

He had bought a round-trip ticket for himself but not for his wife. The clerk thought this odd. When the police checked, the policeman had taken insurance out for his wife's life. He confessed to everything.

40   Front Foot Forward

He was less than one pace from the North Pole when he planted the right foot. His left foot went over the North Pole and was therefore pointing south.

41   DIY Dilema

a) 8 b) 2 c) 35 d) 10 e) 11

42   Mixed Letters

1.   Bathroom, Lounge, Kitchen, Bedroom.

2.   Scissors, Penknife, Tweezers, Corkscrew.

3.   Lorry, Coach, Yacht, Helicopter.

4.   Cabbage, Celery, Beetroot, Pumpkin.

5.   Silk, Linen, Nylon, Cotton.

43   Cocktail Sticks

1.   Take alternate outside cocktail sticks to produce a separate triangle.

2.   Again, remove alternate outside cocktail sticks and overlay them on the others.

3.   A 3-sided pyramid.

4.   Move 2 from any one of A B C or D to form squares at E and F.

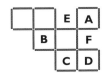

5.   Move both outside cocktail sticks from B or C and complete the squares on E and F. This creates 10 squares of 1 x 1, 4 squares of 2 x 2, and one square of 3 x 3.

ANSWERS

44 Divided Square

A = 99m²	G = 34m²
B = 78m²	H = 9m²
C = 21m²	I = 16m²
D = 77m²	J = 25m²
E = 43m²	K = 41m²
F = 57m²	

45 Blanket

The man was an Indian Chief and he used the blanket for smoke signals to start a battle.

46 Word-Stepping

Penny	Piece
Piece	Meal
Meal	Time
Time	Step
Step	Son
Son	Net
Net	Her
Her	Ring
Ring	Worm
Worm	Screw
Screw	Ball
Ball	Cock
Cock	Pit
Pit	Bath
Bath	Bun

47 Do You Need a Computer?

They are the binary alphabetic positions of the vowels.
A = 1 E = 5 I = 9 O = 15 U = 21

48 The Fire Station Location

Halfway between A and C. This would give a maximum distance of 10 minutes to any town.

49 Letter Grid

1. Pause, lulls, islet, and tests.
2. Aroma, today, amaze, layer.

50 On the Farm

a) Cow – Cattle or   Ox – Oxen
b) Pig – Piglet or   Cat – Kitten
c) Sheep – Sheep or   Swine – Swine

51 What Title Do I Have?

Hope diamond.

52 Confusing Family Relations

1. Alice + 3. Alice's sister is married to Alice's husband's brother. The mother of both men is Alice's mother's sister and she is married to Alice's paternal uncle.

2. So are the other half.

53 Rotations

a) Number moves clockwise by original number.
b) Letter moves clockwise by one less than its numerical position in the alphabet.

A 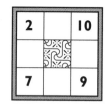   B

54 Missing Vowels

1. Karate, Croquet, Lacrosse, Angling.
2. Ladle, Spatula, Cruet, Cutlery.
3. Greengage, Apricot, Kumquat, Grapefruit.
4. Noodle, Peanut, Pancake, Meringue.
5. Gorilla, Leopard, Reindeer, Porcupine.

55 Moving Water Uphill

Put the pin through the match and pin it to the cork. Strike the match and place on the water so that it floats without getting the match wet. Then put the beaker over the cork and alighted match.

The match burns the oxygen and the water will be drawn into the beaker.

56 Word Squares

a) Valet   b) Chest
   Above   Hello
   Lobes   Elver
   Event   Sleds
   Testy   Torso

57 Card Sharp

a) 19 b) 22 c) 3 d) 45 e) 8

58 Triangles

A

59 Ballet or Opera?

a) 20 b) 12 c) 21 d) 18 e) 122

ANSWERS

60 Pentagons

1

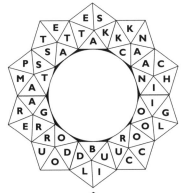

2

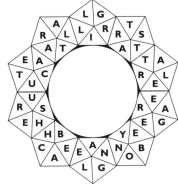

61 Who's Where?
a) Fourth
b) Mr. Brown
c) Ms. Baker
d) Mr. Ridge
e) Fifth

62 Time to Settle Up
Three checks of $20. Penny, Mary, and Claire should give Ann $20.

63 The Striptease Artist
The manager had twisted his back and could not move. The entertainer was only at the bar to earn money to finish a course as an osteopath. The manager had previously used the lady to help him when his back had been twisted before.

64 Who Was I?
Harry Houdini.

65 Scared
The dark shadow was a spider.

66 Incorrect Use of the Calculator
a = 12, b = 36 and c = 3.

67 Spider's Logic
36. Feet positions in web.

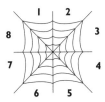

68 Complex Division
381654729

69 Square Meters?
The natural thing to do is to make the field square as the area of a square with the same perimeter as a rectangle will always be larger. The answer, however, is to make it a circle.

π D = circumference of a circle
π D = 3000 = 954.80585 meters or r (radius) = 477.40 meters

π r^2 = area of circle
A = 716104.31 square meters

The area of a square would only be 56,2500 square meters.

70 Decimated
976. Take 2 to the power of 10, which gives you the lowest number above 1000.
$2^{10} = 1024$
Then use the formula 1024 – 2 (1024 – 1000)
= 976

71 Hidden Connections
1. The words contain: Tie, Shoe, Hat, and Belt.
2. The words contain: Cod, Roe, Pike, and Hake.
3. The words contain: Play, Drama, Film, and Show.
4. The words contain: Ape, Rat, Deer, and Elk.
5. The words contain: Lark, Rook, Owl, and Rail.

72 Transmogrification
Plum (Add the B-line to Plum to make Plumb line)

ANSWERS

73 The Diner
Couple 'C'.

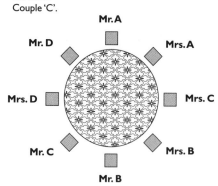

Mr. A
Mr. D — Mrs. A
Mrs. D — Mrs. C
Mr. C — Mrs. B
Mr. B

74 More Matchstick Trickery!
a) Remove any 2 matchsticks from any corner and place as shown.
Remember a block of 4 squares forms a bigger square.
b) Move the remaining match that was left on its own.

ANS A

ANS B

75 Pyramid Letters
1. Penguin, manor, long, toy, tan, you.
2. Too, elm, thaw, the, koala.

76 Old Grandpa!
56 years old. (Seven sons for each brother, and seven brothers: 8 x 7 = 56 total offspring.)

77 Confusing Paper Model
Cut the paper as shown.

Life flap A vertically toward you around the fold line, then twist section B through 180°.

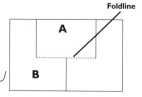

Foldline

A

B

78 Classic Car
An American gallon is equal to about 0.833 imperial (UK) gallons.

79 Unusual Safes
1. 2S on the 2nd row 2nd column.
2. 1A on the inner circle at the 4 o'clock position.
3. S to give Haste, Laser, Miser, and Nasty.
4. G to give Anger, Eager, Bight, and Regal.

80 Cleaning Confusion
a) 15 b) 24 c) 10 d) 34 e) 73

81 Dozy Policemen
It was a sleeping policeman (traffic-calming road bump).

82 Find the Link
1. Trap, lived, regal, and bats all create another word when read backward.
2. Right. All the words on the left have their letters in alphabetical order. The words on the right have their letters in reverse alphabetical order.
3. Hazel. The alphabetical values of the first and last letters total 20.

83 Missing Letters
1.	Scientists	2.	Planets
a)	Planck.	a)	Jupiter.
b)	Euclid.	b)	Pluto.
c)	Edison.	c)	Saturn.

3. Composers
a) Liszt.
b) Dvorak.
c) Mahler.
d) Mozart.

84 Changing Words
1.	Seat	Seam	Team	Tram		
2.	Head	Heal	Teal	Taal	Tail	
3.	Stone	Shone	Shine	Thine	Think	Thick
	Trick	Brick				
4.	White	Whine	Chine	Chink	Clink	Blink
	Blank	Black				
	(also Clank, Clack, Black)					
5.	Here	Hare	Hark	Hank	Hunk	Junk
6.	Fair	Fail	Fall	Fill	Rill	Rile
	Ride					
7.	Write	Writs	Waits	Warts	Wards	Cards
8.	Brown	Brows	Brews	Trews	Trees	
9.	Glass	Class	Clans	Clank	Clink	Chink
	China					
10.	Green	Breen	Bleed	Blend	Bland	Blank
	Black					

ANSWERS

85 But We Need the Beds!

Many patients who are asked to leave a famous hospital in Dublin, on a Saturday, believe that they will die if they are discharged on the Saturday.

(The feeling is strong, and worry caused by this belief leads doctors to permit patients to stay until Monday to get discharged. The reputation of the hospital is held in the highest regard and the superstitions are totally unrelated to work done at the hospital.)

86 Grid Codes

1. Jupiter.
2. California.
3. Pyrenees.
4. Belgium.
5. Richard Gere.

87 The Swiss Deposit Code

1 - 5 - 6 - 2 opens the box
Letter values A=4, B=2, C=5, D=3, E=8, F=1, G=6, H=7, I=9.
DID = IIF (3x9x3) = (9x9x1) etc.

88 The Eleven-Card Con

After the first round, the second player copies every move made by Player 1.

89 What the ??

1. N for November. The letters are the initials of months in the second half of the year.
2. C for Cancer. The letters are the initials of the signs of the zodiac.
3. S for Saturn. The letters are the initials of the planets.
4. A. The letters are the second letters of the days of the week.
5. N. These letters appear along the bottom row of a keyboard.
6. I. The letters are the second letters of the numbers 1 to 9.
7. 5. The number of letters in the star sign's symbol gives the answer, eg. Libra = scales = 6, Aries = ram = 3 etc. Gemini's sign is twins.
8. Jenny has $10 and Paula has $22.
9. On a calculator when the 2 added numbers are seen upside-down. 1089 is actually 6801 and 2568 is actually 8952. 6801 + 8952 = 15753.
10. 4. The number of straight lines in the letters are added together.

90 Party Poser

a) 27 b) 15 c) 6 d) 20 e) 82

91 More Code Wheels

a) Taps.
b) Tabs.
c) Type.
d) Wept.

92 Extinct? I Don't Think So

Any hybrid, such as a jackass, hinney, etc.

93 The Mississippi Gambler

If the customer chose red, he would choose blue.
If the customer chose blue, he would choose yellow.
If the customer chose yellow, he would choose red.
He should win 5 in every 9 rolls

Red vs Blue			Blue vs Yellow		
Red Score	Blue		Blue score	Yellow	
2	3 - 5 - 7	wins	3	6 - 8	wins
4	5 - 7	wins	5	6 - 8	wins
9	Nil	wins	7	- 8	wins

Yellow vs Red		
Yellow Score	Red	
1	2 - 4 - 9	wins
6	9	wins
8	9	wins

94 Hidden Words

1. Flowers: Iris, Pansy, Orchid, and Aster.
2. Rivers: Seine, Rhone, Congo, Ganges, and Wear.
3. Professions: Chef, Miner, Driver, and Joiner.
4. Insects: Flea, Slug, Wasp, and Locust.
5. Foods: Bran, Flan, Bun, and Toast.

95 Devious Clues

1. Opal ring. The first 2 letters of the woman's name followed by the first 2 letters of the man's name give the gem.
2. Gary. The last letter of each capitalized word gives the names.
3. Ann. The last letter of the boys' names gives the girl's name.
4. Australia. The first letter of the first location gives the first letter of the presenter's name. The second letter of the second location gives the second letter of the name and so on.
5. Fruit. The first letter of the name gives the first letter of the product. The second letter of the town gives the second letter, and so on.
6. December. All the words contain one vowel repeated and no other.
7. Coffee. Carol likes items that begin with the letters of her name.

96 Mysteries of Time

Yes. His 18th birthday was yesterday, New Year's Eve. He was speaking on New Year's Day, so he will indeed have another birthday in the current year.

97 Link Words

a) Horse.
b) Glass.
c) Show.
d) Time.

98 The Warehouse Sale

40 - 100 - 120 (40 @ 40c = $16) + (100 @ $1 = $100) + (120 @ $1.20 = $144)
    Total = $260

99 Links?

1) 1536. First 2 digits x second 2 digits form next number.
2) 108. Multiply 2 outer digits of first number to form outer digits of next number. Multiply 2 inner digits of first number to form 2 inner digits of second number.
3) 27. The first 2 digits less second 2 digits form next box. Then first digit less second digit = third box.

100 Waiting Game

a) Second
b) Mr. Sharp
c) Mr. Jones
d) Ms. Fielding
e) Fifth

101 Logical Deductions of Who or What am I?

1. **What am I?**
   A stallion
2. **Who am I?**
   Moses
3. **What am I?**
   A calculating machine : abacus/calculator
4. **What am I?**
   A seahorse
5. **What am I?**
   Samba

102 The Rail Workers

They were working on a long bridge with no spare room at the side of the tracks. It was a much shorter distance to the end of the bridge where the train was coming from. They were able to get to the end and jump to one side.

103 Triangles

... 15, 6, 1. The sequence is the seventh row of the additive number triangle.

104 Conveyor-Belt Conundrum

a) Fruit juice
b) Soup
c) Apples
d) Second
e) Biscuits

105 Merged Words

1. Partridge, Chaffinch, Ptarmigan.
2. Squirrel, Mongoose, Chipmunk.

106 Children's Age

Children 2 - 5 - 8 - 11 - 14 - 17 - 20 - 23 - 26 : Father 48

107 Handyman

They were originally 3-inch tiles, and now he would have to use
324 + 250 = 574 tiles.

108 Poisonous Insect

With the lights in the house dimmed she held a glass to the wall covering the hole and shone a beam of light through the glass into the hole. When the insect moved out of the hole and into the glass she slid a card over the end of the glass.

109 Who Am I?

Sir Arthur Conan Doyle.

110 Unlucky Sailor

He got an albatross at golf: 3 under par for a given hole. The mere mention of albatross on some boats is unlucky, so he was thrown overboard.

111 Japanese Door Sign

It was on a glass door in two shades. From one side it said PUSH and from the other it said PULL.

112 Equal Shapes

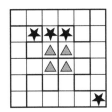

113 Take a Second Look

Second letters of the numbers 1 to 7. Missing letter is E.

114 Lost Elevator

a) Ted
b) Claudia
c) Sophie
d) Joanne
e) Mark

115 Code Wheels

a) Star
b) Clip
c) Scab
d) Drab

116 Skyscraper

He was cleaning the inside of the windows.

117 The Fairground Game

a) 15
b) 3
c) 24
d) 20
e) 12

118 Number Links

1. Brontosaurus.

B		R		O	
S	R	R	T	T	N
B		R	U		A
B		U		A	
U	O	O	A	A	T
T		S		O	
T		S		O	
R	O	O	B	B	O
U		A		S	

2. San Francisco

S		A		N	
O	E	F	M	N	F
B		H		H	
A		I		I	
C	I	J	F	G	R
L		D		A	
M		E		B	
S	M	N	E	F	A
I		C		N	

119 The Panel Game

A

Truth Table	A	B	C
I have it	T		L
'A' does not have it		L	
'A' nor 'C' has it	L		
'B' does not have it	T	T	T
'C' does not have it		T	T

T = TRUE
L = LIE

120 Reader Riddle

a) 76 b) 24 c) 104 d) 47 e) 227

121 What Am I?

1. Gravy, then grave, grate, grape, and graph.
2. Store, then stork, stark, start, and smart.

122 The Train Driver

You are the train driver, so will see whatever color eyes you have.

ANSWERS

123 The Gearbox

26 times + 240° clockwise.

124 Columns

1. Pear, Plum and Lime.
   People
   Eyelid
   Autumn
   Rhymes

2. Italy and China.
   Infancy
   Thought
   Angelic
   Lenient
   Yashmak

3. McLaren and Ferrari.
   Mindful
   Closest
   Liberty
   Admiral
   Radiant
   Emperor
   Nourish

4. Trifle and Toffee.
   Talent
   Robust
   Inform
   Fulfil
   Ladder
   Empire

125 The Car Problem

The bottom half of the wheels.

126 Magic Squares with Dominoes

3 - 6	2 - 6	2 - 5	3 - 3	0 - 0
3 - 5	0 - 2	0 - 1	4 - 6	4 - 5
1 - 1	5 - 6	5 - 5	1 - 3	1 - 2
2 - 4	1 - 4	2 - 2	0 - 3	6 - 6
2 - 3	0 - 4	4 - 4	3 - 4	1 - 5

127 Currencies

Yen, Pence, Pound, Ecu, Cent, Mark, Franc, Peso, Rand, Krona, Dollar, Peseta, Shekel.

128 Secret Messages

Ja COB-ALT	(-)	=	Cobalt
AL ter	Al	=	Aluminium
AU gust's	Au	=	Gold
GE rmany	Ge	=	Germanium
unfa IR-ON	(-)	=	Iron
o NE	Ne	=	Neon
NI le	Ni	=	Nickle

129 Circular Crossword

East Star.

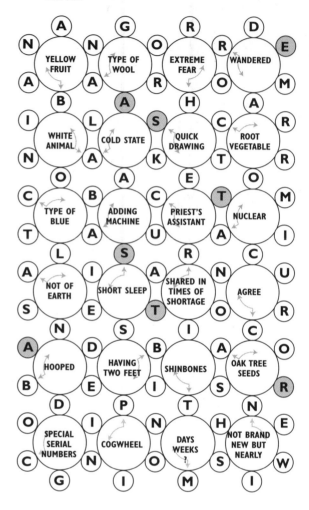